Science Studies Meets Colonialism

T0188457

For Zara

Science Studies Meets Colonialism

Amit Prasad

polity

First published in 2023 by Polity Press

Polity Press
65 Bridge Street
Cambridge CB2 1UR, UK

Polity Press
111 River Street
Hoboken, NJ 07030, USA

ISBN-13: 978-1-5095-4441-7
ISBN-13: 978-1-5095-4442-4(pb)

A catalogue record for this book is available from the British Library.

Library of Congress Control Number: 2022934667

Typeset in 11 on 13pt Sabon LT Pro
by Cheshire Typesetting Ltd, Cuddington, Cheshire
Printed and bound in the UK by CPI Group (UK) Ltd, Croydon

The publisher has used its best endeavours to ensure that the URLs for external websites referred to in this book are correct and active at the time of going to press. However, the publisher has no responsibility for the websites and can make no guarantee that a site will remain live or that the content is or will remain appropriate.

For further information on Polity, visit our website:
politybooks.com

Contents

Preface

Some years ago, during a colloquium on my first book, *Imperial Technoscience: Transnational Histories of MRI in the United States, Britain, and India* (Prasad 2014), the person commentating on the book suggested that perhaps I should have focused on aspects of caste and its relationship to scientific practices in India. I was surprised but also realized that it was not an unusual suggestion. I remembered, many years before that colloquium, when I was planning to go to the United States or the United Kingdom to do my PhD, I was twice a finalist for the Inlaks Scholarship and I was informed through one of my mentors that the selection committee liked my profile but they thought that I should work on a different project. My mentor laughed and said they need to get over wanting Indians to work only on caste and tribe. I was also reminded of a joke that some of my Indian-origin academic friends have shared at times – if we critique the West, how long does it take for a western scholar, particularly those of British origin, to remind us about caste hierarchies in India? Interestingly, just a few days before the above-mentioned colloquium, I had been told by one of my colleagues at my former academic home (when I was highlighting the impact of liberal white racism in the department) that being a foreigner I did not understand the racial context of the United States.

I mention my experience to situate the Orientalized entrenchments that are commonly, albeit inadvertently, envisioned for non-western scholars in academia. Suggestions like the one that I mention are reflective of "othering" that seeks to position non-western scholars in particular boxes that tie the expertise of the scholar to his/her/their identity. Let me clarify. It is not that we do not need to study the relationship of caste and scientific practices in India – I can bet the scientific institutions in India are still almost exclusively dominated by so-called upper-caste people (and those too largely male). I do not, however, discuss caste hierarchies in the practice of science in India in my earlier book. My aim was to problematize the international hierarchies in the invention, development, and marketing of magnetic resonance imaging (MRI). Specifically, I showed how, even though dualist hierarchies between the West and the non-West and across different nations bear on techno-scientific research, yet one can map the genealogies of entangled exchanges that cut across these boundaries (Prasad 2014).

I discuss and problematize my experience neither to defend my earlier work nor to blunt possible critiques of this book, which also does not discuss the relationship between caste and scientific research in India. My concern is the opposite. My studies, like those of anyone else, are situated in my various intersectional social locations, namely caste, class, gender, non-western, postcolonial, and so on. I may not even be self-reflexively aware of the impact of these locations on my studies. Hence, critique is necessary. My claim is simply that the study of the relationship between caste hierarchy and scientific practice requires careful analysis, and it is not useful to see it as separate or separable from analysis of the relationship between science and colonialism. In the intellectual and lay discussions on science and on the role of scientists in India, we often slip into idealized nostalgia. For example, Shiv Viswanathan, a leading sociologist of science from India, wrote, "Science too added to the colours of nationalism making Indian science both nationalist and cosmopolitan."[1] The fact that this cosmopolitanism was and continues to be under male and "upper-caste" control is forgotten. Itty Abraham, who is more circumspect in highlighting the hierarchies within scientific research in India (Abraham 1998), as Abha Sur (2002) points out, misses the caste and associated social justice dimensions that have undergirded, and continue to undergird, the hierarchies in the scientific field in India.[2] More broadly, Sur argues that the nationalist imagination, while it "strove to puncture the myth of British superiority ... depended upon the creation of new myths and new heroes," which in turn were inextricably tied to the existing "social and political configurations" in India, including that of caste hierarchies (Sur 1999: 48–9). Nevertheless, it is not straightforward to show the role of caste hierarchy in scientific practice because the discourse of modern science in India (and other non-western countries) is a derivative discourse and as such is inextricably tied to European colonialism and Euro-/West centrism. That simply means that we need more studies to investigate the complex configurations of science in different locales. This should also necessarily entail critiques of existing historical and sociological studies of sciences (including mine). Of course, critiques, like any other analysis, can remain complicit in colonial or other dominant discourses, in which case we need to expose that complicity. And if we come to the conclusion that "critique has run out of steam," that simply means it is time for us to refuel critique, rather than abandoning critique altogether.

Acknowledgments

First and foremost, I wish to thank Jonathan Skerrett of Polity Press for persisting with his interest in my writing this book and for providing support throughout the process. I was not sure of writing such a book. I thought there was not much to add to what has already been written on post/decolonial science studies. I must admit I was also anxious about critically engaging with some of the science and technology studies (STS) scholars whose work I not only admire, but who have been and continue to be my mentors. In fact, in my previous book, *Imperial Technoscience* (Prasad 2014), I barely discussed postcolonial science studies for the same reason. I still vividly remember Sandra Harding's comment in one of the first STS workshops I had attended (which was at Virginia Tech) – "when we talk about science, they call it feminist squabbles, and when they do it, they call it criticism and growth of knowledge." I am also personally aware of the deep commitment of Warwick Anderson in pursuing postcolonial science studies within the field of STS. Hence I thought if I engaged with postcolonial science studies, I would end up critiquing their positions and that may not be good for the field.

However, as I started writing this book, I also felt that not being able to highlight the problems with postcolonial science studies (and more broadly STS) was having a paralyzing impact on me. I was not happy with the way engagements with colonialism (and in a sense with globalization as well) were taking place within STS, and my response was to avoid participating in, for example, the 4S conferences. In that sense writing this book has been liberatory at a personal level. I must, nonetheless, clarify that my interest is not – as one reviewer of the book put it – "a root and branch unmasking of these more than minor vestiges of the colonial mindset within the dark heart of science studies." I have written this book because I think it is now time for post/decolonial interventions within STS to move further or be abandoned; however, abandoning the post/decolonial debates within STS also runs the risk of critically jeopardizing STS interest in moving beyond its Euro-American-Australian backyard and becoming truly global.

I wish to thank the three anonymous reviewers whose comments, criticisms, and suggestions were extremely helpful. In particular, I cannot thank enough one of the reviewers who took immense care to point out even minor grammatical and analytical errors, including those in the footnotes. Even the reviewer who did not like the manuscript provided very useful suggestions and criticisms that helped me to more clearly articulate my concerns. The analysis presented in the book owes to the many discussions that I have had in the past with STS as well as non-STS scholars. If I have forgotten to cite some of these people's work, I apologize in advance. I wish to especially thank Projit Mukharji, Alexandra Hofmänner, Dwaipan Banerjee, Bharat Venkat, Warwick Anderson, Banu Subramaniam, Gabriela Soto Laveaga, and Itty Abraham with who I was directly or indirectly intellectually engaging with the issue of colonialism and sciences while writing the book. I wish to particularly thank Itty Abraham, Dwaipan Banerjee, and Mary Macdonald for reading drafts of some of the chapters. I would also like to thank Karina Jákupsdóttir for her editorial support and Gail Ferguson for carefully copy-editing the book manuscript. Lastly, and most importantly, I wish to thank Srirupa and Zara for all their love and support.

Introduction

Genealogies of Colonialism in Postcolonial Times

> China is mounting a serious challenge to the United
> States for global leadership in technology and innovation
> . . . In fact, for the first time Chinese investment in R&D
> will surpass US investment next year.
>
> Representative Robin Kelly, ranking member,
> Subcommittee on Information Technology,
> US Congress, 2019

> East India Company returns after 135-year absence.
>
> British Broadcasting Corporation (BBC),
> August 13, 2010

On September 26, 2018, the second session of the hearing before the Subcommittee on Information Technology of the United States (US) House of Representatives started with the following statement by Representative Will Hurd, the Chairman of the Subcommittee: "For more than 40 years, the US has encouraged China to develop its own economy and take its place alongside the US as a central and responsible player on the world stage, but China does not want to join us. *They want to replace us*" (emphasis added).[1] The hearing, which was unequivocally titled, "Countering China: Ensuring America Remains the World Leader in Advanced Technologies and Innovation," is a testament

1

to multilayered concerns in the United States, and more broadly in the western countries, about the rise of China as another global "superpower."

The US Congress has not been alone in voicing such concerns. Technological competition between the United States and China has been widely reported in the media and extensively discussed by policy experts. In fact, "policy communities on all sides appear to agree that old-fashioned trade frictions are really just an appetizer ahead of the main course: a looming 'technology war'" (Lim 2019: 8). In the United States, at a time when political division has guided dealing with even the COVID-19 pandemic, concern with China's fast technological growth has brought together the Republicans and the Democrats, and "China has emerged as one of the few issues on which even Democrats agree that [former] President Trump had some valid points."[2] This rare bipartisanship in the US Congress was visible when "The United States Innovation and Competition Act," also known as the "Endless Frontier Act," was passed in the US Senate on June 8, 2021 with slightly more than a two-thirds majority.[3] The White House in its statement on the passing of this Act in the Senate was categorical: "*We are in a competition to win the twenty-first century, and the starting gun has gone off*," adding, "America must maintain its position as the most innovative and productive nation on Earth" (emphasis added).[4]

The attitude towards China's technological growth has definitely changed in the last decade. In 2010, the *Harvard Business Review*, while stating that "almost unnoticed China has been moving towards a new stage of development . . . to a sophisticated high-tech" economy, had, for example, also noted: "Even as China moves up the ranks of economic superpowers, many discount these recent milestones. They don't believe China will become richer than the US."[5] Four years later, in 2014, another article published in the *Harvard Business Review*, after posing the question "whether China has a good institutional framework for innovation," had categorically stated, "our answer at present is no." Indeed, as the title of the article made clear, the aim of the authors was to explain "Why China Can't Innovate" (Abrami, Kirby, and McFarlan 2014).

A significant reason for the shift in attitude with regard to China is the concern that Representative Hurd had put at the hearing before the US House of Representatives: "They want to replace us." Although concern with the "rise of China" may be understandable, at least in terms of geopolitical strategy, the language that is being used to describe that concern often reflects a colonial and racial genealogy. "You will not replace us" has, for example, become a rallying cry for white supremacist groups in European countries as well as in the United States, as was strikingly evident in Charlottesville, Virginia in 2017.[6] However, the colonial genealogy often gets masked in the language of liberty, freedom, and democracy, and, ironically, the genealogical links of liberalism and empire are forgotten in the process (see, e.g., Mehta 1999).

A recent report of the National Bureau of Asian Research, while emphasizing that the "expectations and assumptions" regarding engagement with China "were widely shared on both sides of the aisle in American politics," for example, quotes the former President George W. Bush: "Economic freedom creates habits of liberty. And habits of liberty create expectations of democracy ... Trade freely with China, and time is on our side" (as quoted in Boustany and Friedberg 2019: 4). This goal, according to the Report, was based on a long-cherished western and American principle "that trade promotes both national welfare and international understanding ... and that the spread of democracy will eventually usher in an era of lasting world peace" (Boustany and

Friedberg 2019: 5). The Report, thus, argues that the emerging crisis in the US–China relationship is because "whereas US leaders have historically been believers in liberal democracy and free-market capitalism, China's rulers are adherents of what can be best described as 'mercantilist Leninism'" (Boustany and Friedberg 2019: 7).

The discursive framing of the United States' technological competition with China does not, however, unambiguously express racial, colonial, or western hierarchy. It commonly slips into the biopolitical template of colonial/racial stereotype that, as Homi Bhabha (1994) argued, is undergirded by ambivalence. The statement of Representative Hurd that I quoted earlier, for example, re-presents, to draw on Homi Bhabha, the ambivalent desire to constitute and contain China and Chinese people through the colonial discourse of mimicry – "to be Anglicized is *emphatically* not to be English" (Bhabha 1994: 125). China was welcomed to develop and "take its place alongside the US," but the boundary of this desired mimicry, just like the colonial times, could not be crossed – the mimic was always "not quite/not white." However, at present, in the context of a radical shift in geopolitics and global economy, often, the non-western "others" are not content with "mimicry" of the West, hence the added anxiety, "they want to replace us." The postcolonial/colonial discourse through which the US–China tech war is being framed has, thus, become doubly ambivalent. On the one hand, it reflects "projection and introjection, metaphoric ["narcissistic"] and metonymic ["aggressive"] strategies" (1994: 117) that guided the colonial discourse of mimicry and, on the other, it calls for the western "self" to stand against and catch up with the "other." These ambivalent discursive strategies were evident at the hearing of the Subcommittee on Information Technology that I have been discussing and are also commonly articulated in a number of contexts and by a range of western actors.

Representative Hurd, while highlighting the efforts of China to "become the world leader in manufacturing," for example, emphasized, "we cannot lose sight of the many abuses the oppressive Chinese government perpetuates on its citizens." He goes on to claim that the "Communist Party that runs the Chinese government is an oppressive regime with an abysmal human rights record." And he presents the contrast through the United States. "As Americans, we believe everyone has a right to life, liberty, and the pursuit of happiness." The earlier-expressed concern with reference to China's "aim to modernize the Chinese economy" soon slips into a colonial desire for the Chinese people: "The millions of men, women, and children living under the oppressive Chinese regime in China deserve better from their government." The very next sentence shifts to presenting "the East" as an impending threat – "Our nation faces a great challenge rising from the East." A similar trope was also used by Herbert Butterfield to explain the impact of "Asiatics" on medieval Europe prior to the Scientific Revolution, as I show in chapter 2.[7] At present, the temporal structure of this threat has shifted and is shown as directed not at the medieval past ("dark ages") of Europe/the West but at its future.

Such postcolonial/colonial ambivalence is present even in more nuanced and critical analyses of technological competition between the United States and China. An article exploring the "legal regimes recently developed in both countries to wage the tech war," for example, laments: "We fear that ironically, the rule of law necessary to maintain continued vibrancy in US high-tech sectors is being compromised by some of the very actions ostensibly taken to protect these sectors from malign foreign influence." The footnote accompanying the above quote states, "China has managed thus far to develop economically without a robust rule of law" (Milhaupt and Callahan 2021). Such broad claims about the "rule of law" constitutes China as the stereotypical "other" – harking back to the old colonial trope of the non-West as the "zone of lawlessness" and the West as the "rule of law" (Benton 2010). Moreover, what is forgotten in the common European/western self-presentation as the repository of the "rule of law" is an imperial genealogy of western understanding of law: "land appropriation [through various means including colonialism]," to which the Europeans later added "sea appropriations," has been "the primeval act in founding law" (Schmitt 2006: 45). In the words of Achille Mbembe, "whenever Europe referred to the principle of liberty in relation to the World-outside, what was really meant was an absence of law and organized civil society, which authorized the free and unscrupulous use of force" (Mbembe 2017: 59).

The very fact that the rising competitiveness of China in high-end technologies is seen as a threat to the *West* is itself telling. LSE Ideas, the foreign policy think tank of the London School of Economics, for example, published a series of essays online that were titled "Protect, Constrain, Contest" to present "[a]pproaches for coordinated transatlantic economic and technological competition with China." Peter Watkins, in the introduction to this series, starts his essay thus: "Over the past year, there has been a growing realization in the *traditional 'West'* – including the United States, the European Union and the United Kingdom – of the *challenge posed by China*" (emphasis added). Similar framing of the China threat, as I show in chapter 1, has been frequently articulated in the context of the failure of the "traditional West" in the management of the COVID pandemic and China's diplomacy to gain global influence. The COVID pandemic has indeed added urgency to "western" anxieties in relation to China and the non-West, as Watkins highlights: "The academic consensus in the early months of the pandemic was that it would accelerate existing geo-economic and geopolitical trends (including the shift in economic power from the 'West' to the Indo-Pacific)." The concern with China in particular, according to Watkins, is because China is "not simply an economic competitor," even though it is "not yet a military challenge in the same way as Russia . . . during the Cold War." He goes on to suggest "precautionary steps to protect the West's own vital economic interests which may sit uneasily with a fundamentally free-market approach."[8]

This coming together of the "traditional West" in the name of the "free market" in relation to China has another colonial genealogy – the Opium Wars (1839–1842 and 1856–1860), which in China are portrayed as marking the start of China's "century of humiliation" that ended with the Sino-Japanese war in 1945. The Office of the Historian of the United States Department of State calls "the First Opium War" and the Treaty of Wangxia, which was signed between the United States and China in 1844, as "the Opening of China" and lists it as one of the "Milestones in the History of US Foreign Relations."[9] Interestingly, the treaties that the western countries had signed after the defeat of China rarely mentioned opium – whose trade was forced on China – and resulted in a shift in favor of the western countries. In the forked colonial articulation of the treaties (that western countries imposed on the Chinese emperor) the sanitized language of protection of property and rights for free trade at the Chinese ports made opium only a spectral presence. The Treaty of Nanjing that was signed between Britain and China in 1842, for example, only mentions opium once, in Article IV of the Treaty: "The Emperor of China agrees to pay the sum of Six Millions of Dollars as the value of Opium which was delivered up at Canton in the month of March 1839, as a Ransom for the lives of Her Britannic Majesty's Superintendent and Subjects, who had been imprisoned and threatened with death by the Chinese high Officers."[10]

Although the implications of the "Century of Humiliation" are known in the policy circles of western countries, it remains "the most misunderstood and least discussed element" in relation to China (Billy Wireman, as quoted in Wang 2014a: 2). In 2011, a China analyst, for example, gave testimony before the US–China Economic and Security Review Commission Hearing on "China's Narratives Regarding National Security Policy," and she specifically focused on the "Century of Humiliation." She went on to suggest that the "policymakers must assess the extent to which the founding narrative of the Century of Humiliation, and the values and aspirations derived from it, can be used today to persuade China's leaders to move down an accommodating peaceful path."[11] Nonetheless, recent responses of many policymakers in the western countries, even while seeking "broader western relationship with China," not only ignore any discussion of the history of colonialism that underlies the discourse of the "Century of Humiliation," they often articulate the opposition to China through colonial tropes, albeit not in the older colonial language. Moreover, these responses commonly deploy the friend-versus-enemy dialectic that Carl Schmitt, a critic of liberal polity and a prominent ideologue of the Nazis, argued was the basis of the "political" (see also Mouffe 1997; Schmitt 2007).[12] LSE Ideas, the foreign policy think tank of the London School of Economics, for example, presents the desired response of the "West" to China through the categories of "protect," "constrain," and "contest," wherein contest is defined as follows "Sustaining the Western position in the face of the China challenge cannot be a purely defensive game."[13]

Failure to engage with the genealogies of colonialism constrains not only the West but also China within the Orientalist "imaginative geography" that intensifies the difference and distance between the "self" and the "other" (Said 1979). The discourse of the "Century of Humiliation" is an illustrative example in this regard. Though the "national humiliation discourse" does "not receive much attention in Western analysis," as William Callahan found to his surprise, "there are textbooks, novels, museums, songs, and parks devoted to commemorating national humiliation in China" (Callahan 2004: 199). Callahan goes on to "argue that the master narrative of modern Chinese history is the discourse of the century of national humiliation" (2004: 204). Indeed, as Zheng Wang has argued, "the discourse of national humiliation is an integral part of the construction of national identity and nation building" in China (Wang 2014a: 2) that, according to him, constitutes China's "CMT (Chosenness–Myths–Trauma) complex." Wang analyzes "the role of historical memory in China's nation building" (Wang 2014b: 10) and shows how, since President Xi Jinping took office in 2012, "he [Jinping] has promoted the concept . . . [of] 'the Chinese Dream'" "to realize the great rejuvenation of the Chinese nation . . . [as] the greatest dream for the Chinese nation in modern history" (Wang 2014a: 1). As a result, far from the commonly expressed concern in relation to Chinese citizens (that, for example, Representative Hurd expressed), it becomes "a puzzle for China watchers to try to understand why Chinese youths are increasingly nationalistic based on events that took place in [a] distant past" (Wang 2014b: 10). More broadly, because the historical/colonial context of the "Chinese Dream" is not known outside China, except "indirectly and tangentially," we are left with "perception gaps" (Wang 2014a, 2014b) in dealing with perhaps the most important geopolitical issue of our time – a looming West-versus-China technology war.

I have discussed the looming West-versus-China technology war to show how mapping of the genealogies of colonialism allows us to better understand the history of the present. To situate European colonialism simply in the distant past is to ignore that individuals as well as societies/nations are discursively constituted and imagined through historical experiences. My concern, similar to that of Michel Foucault, is "not to demonstrate that the past actively exists in the present" but to explore how the genealogies of the past continue to "animate the present" (Foucault 1977a: 146). I follow Foucault's genealogical approach, which instead of seeking an origin (e.g., colonial origin of present-day events) or attempting to "restore an unbroken continuity" to the past (Foucault 1977a) seeks "to show 'descent' and 'emergence' and how the contingencies of these continue to shape the present" (Garland 2014: 371).[14]

The term "emergence" implies contingency and open-endedness and has been commonly used in the field of science and technology studies or STS (see, e.g., Pickering 1995).[15] My use of the term retains STS concern with contingency but is deployed in the Foucauldian sense to argue that "[e]mergence is . . . produced through a particular stage of forces" (Foucault 1977a: 148–9). That is to say, "entities do not exist in-themselves . . . they exist . . . as functions of other relations" (Wilson 1995: 160), and their particular articulations are historically contingent. As such, the "different points of emergence" are not simply successive configurations of identical meaning," i.e., repetitions of the past; rather, they represent "substitutions, displacements, disguised conquests, and systematic reversals" that reconfigure the remnants of the past (Foucault 1977a: 151). This book explores the role of European colonialism in animating and constituting the present similarly. Genealogies of colonialism, if we are willing to look closely, are evident in a wide number of contexts and practices across the globe, albeit their articulations are different and often also ambivalent and contradictory.

The second quote at the start of this introductory chapter – "East India Company returns after 135-year absence" – is another reflection of such an ambivalent and contradictory articulation. The East India Company, as we know, not only colonized most of India before India was brought under the British crown in 1858, but was also the central actor in the production of opium (in India) and its sale (in China) that led to the Opium Wars and shifted the trade imbalance of western countries with China. After the Opium Wars with China and the 1857 revolt in India, the East India Company became dysfunctional for all practical purposes.

The East India Company has a unique place in global history because its activities in the eighteenth and the nineteenth centuries not only overlapped with the interests of the British nation-state, the Company was a partner in British commercial and war efforts that established Britain as the largest and most powerful colonial power. In the words of Edmund Burke, who played a pivotal role in the impeachment of Warren Hastings, the first Governor General of Bengal after the East India Company's takeover, the Company was "a state in the guise of a merchant" (as quoted in Dalrymple 2019: 3). "By 1805 the Company's three Indian armies numbered almost 200,000 men and they were increasingly maintained, via subsidiary alliances, at the expense of Indian rulers such as the Nawab Vizier of Awadh" (Bowen 2006: 47). These Company troops were commonly deployed by Britain for its ever-growing colonial wars. For example, in the late eighteenth and the early nineteenth centuries, "Company troops and ships were sent to Ceylon (1795), Malacca and the Moluccas (1795), the Cape (1795 and 1806), Egypt (1801), Mauritius (1810), and Java (1811)" (2006: 47). The Company, along with troops and financial support, also provided resources from India for British war efforts, e.g., saltpeter and food grains. The role of the East India Company was recognized even by the Royal Society that elected Lord Clive, who had done no scientific research in India or Britain, as a Fellow of the Royal Society in 1768 "as *a reward for his exploits in India*" (Home 2002: 313; emphasis added).[16]

The imaginary and historiography of the East India Company, in a sense, has mimicked the vicissitudes of empire's role in defining British identity that continues to the present. On the one hand, the East India Company is still hailed as "one of the world's greatest trading companies" (Ferguson 2002: 15) and, on the other, the Company's role in the "subjugation and plunder of vast tracts of southern Asia" has been described as "the supreme act of corporate violence in the world" (Dalrymple 2019: xxvii). Even the critiques of the East India Company have remained complicit with the "imagining" of the role of the British Empire. As Nicholas Dirks showed, the critiques of the East India company that were presented, for example, at the time of the impeachment of Warren Hastings (which included those by Edmund Burke, who I quoted earlier), served the purpose of deflecting "the scandal of empire" and recasting the British Empire's role as a civilizing mission (Dirks 2006).

Although the East India Company has continued to have spectral and real presence in a wide variety of ways, the revival of the East India Company is nevertheless puzzling.[17] The revival becomes all the more puzzling when we come to know that the East India Company was bought by an Indian businessman, Sanjiv Mehta, for this purpose.[18] In 2005, Mehta, who had "set up business in London in the 1980s, "bought the entire company, which gave him the rights to trade using its name, and its coat of arms as a trademark."[19] The *Guardian*'s report on this acquisition, from which I quoted above, was written by an Indian freelance journalist (Vidhi Doshi, who published the report from Mumbai) and titled: "How the East India Company Became a Weapon to Challenge UK's Colonial Past." It is not that Mehta did not know about the East India Company's history. He had "learned about it as a schoolboy in newly independent India" and his family, as Mehta shared with the reporter, "had fought alongside Gandhi for India's independence." Moreover, from 2003, when he decided to buy the East India Company, Mehta spent several years "every weekend and spare hour studying the company's history in the British Library and archive of the Victoria and Albert Museum" to know more about the Company's history. "His studies," the *Guardian* report elaborates, "inspired trips around the world." Mehta shared with the reporter that he "wanted to go everywhere that the East India Company had ever set foot."[20]

How are we to read and interpret the acquisition of the East India Company by an Indian businessman? Is Mehta's acquisition, as the *Guardian* report claims, "a weapon to challenge UK's colonial past"? Mehta, according to a BBC report, "dismissed fears that the reappearance of a company long associated with colonialism would open old wounds."[21] He claimed: "It's a disproportionate joy, [I have received] more than 15,000 emails [of support] from various Indians across India, even from Barbados to Fiji to Canada to Boston."[22] The East India Company in its new avatar, which is as a luxury brand, far from challenging the British colonial past, celebrates the Company's colonial achievements. Mehta's "The East India Company" describes itself thus: "When you hear our name you may already have a sense of who we are. *Deep within the world's sub-consciousness is an awareness of the East India Company, powerful pictures of who we are*" (emphasis added).[23] The description goes on: "The East India Company made a wide range of elusive, exclusive and exotic ingredients familiar, affordable and available to the world … *Today we continue to develop and market unique and innovative products that breathe life into the history of The Company*" (emphasis added).

Rather than challenging British colonialism and the plunder and destruction of India by the old East India Company, the Mehta-owned Company extols the imperial power and whitewashes colonial appropriations through the language of making available "elusive, exclusive and exotic ingredients." The description of the Company's philosophy states, "Over three hundred years ago, the East India Company's pioneering merchants were the leading traders of luxury goods of their time, forging trade routes across the tumultuous oceans between the East and the West."[24] It adds: "Today, we *continue to stand* for quality and innovation in each and every area of trade" (emphasis added). In a section titled "Pioneering Spirit," a photo of a British/ European woman using a hand telescope standing next to a British/European man, who is sitting, both possibly looking into the horizon of the "exotic" lands that were colonized by the British/Europeans, accompanies the following statement: "At the heart of it all, we are adventurous, honourable merchants; traversing the oceans to forge new connections, to set up trading relationships, to discover the exotic and elusive."[25]

In short, Mehta's East India Company mimics the imperial/colonial image of the older East India Company. If there could be any doubt about that, it is dispelled in another description on the Company's website that is accompanied with the images of the old East India Company ships sailing in the ocean and is titled "Legacy":

Over the centuries The East India Company unleashed great forces and its activities changed the tastes and thinking. Its officers were adventurers and merchants.

They set sail in a spirit of expansion to establish new trading routes and factories, to discover and bring back new
products and to break down the barriers of the world,
driven by passion, determination and enthusiasm.[26]

The description goes on to praise the historical role of
the Company and then briefly adds: "However, some of
the original Company's and Officers' actions, whilst at
the time widely considered by society to be acceptable
consequences of growth and commerce, today of course
are rightly viewed in a very different light." The web
description passingly mentions the slave trade before
ending with the following words that celebrate the
genealogical link of the present Company with its past
(colonial) history: "Today, the values and behaviours of
the modern reinvented company and brand retain the
pioneering spirit and curiosity of the original Company
men, but in other ways are completely the opposite to
those of the past."[27]

The takeover of the East India Company by Sanjiv
Mehta was one of several takeovers of British and
European companies by Indian business groups in the
first decade of the new millennium, e.g., Tata bought
Jaguar/Land Rover and Corus Steel. However, there
is a key difference in this case. Mehta bought, in business terms, a dysfunctional company and one that was
directly associated with colonialism during the time
it prospered. The East India Company, when Mehta
bought the company, only offered symbolic capital (and
that too because of the association of the Company with
British colonialism), which Mehta, as the web description of the Company illustrates, sought to translate to
economic capital.

The only way the East India Company in its new
avatar challenges British colonialism is by making a
mockery of the racial and colonial divide that defined
colonial mimicry. "Not quite/not white," which Bhabha

described as the discursive strategy of colonial mimicry (Bhabha 1994), becomes "not white/but quite." However, in doing so, Mehta's East India Company whitewashes the wide-ranging and enduring impact of British colonialism, particularly through the Company. This is yet another genealogy of European colonialism that we see at present. Its ambivalent and contradictory articulation rather than transcending colonial mimicry simply presents a mutation that would, most likely, not limit or end the "narcissism and paranoia that repeat furiously, uncontrollably" within the colonial discourse of mimicry (Bhabha 1994: 131).

Science studies meets colonialism

The owl of Minerva spreads its wings only with the falling of the dusk.
G. W. F. Hegel. *Philosophy of Right*. 1821

The staggering number of COVID infections and deaths in the western countries (see chapter 1) and the swift fall of Afghanistan to the Taliban forces following the withdrawal of the United States after 20 years of occupation, which was supported by several other western countries, has left the image of the West in tatters.

That does not, however, mean an end of colonial ideologies and practices. Indeed, as I hope to show through this book, the persistence of genealogies of colonialism, to draw on Pierre Bourdieu's (1977) concepts, has resulted in a crisis wherein colonial orthodoxies are still present across the globe, but heterodoxies that are contesting colonial ideologies and practices are also strikingly visible at the local and the global levels. The confrontations of orthodoxies and heterodoxies, as I showed with the examples of the US/West versus China technological war and the revival of the East India

Company, are complex and often muddle the West-centered hierarchy.

Let me, however, clarify what I mean by colonialism and its genealogies. I am certainly not suggesting that European colonialism never ended. It did formally end in the second half of the twentieth century when the erstwhile colonized countries gained independence. My concern is with the colonial tropes, strategies, and stereotypes that often undergird global, national, and local discourses, colonial law and jurisprudence that have continued to persist in postcolonial times (e.g., the various iterations of laws against homosexuality in erstwhile colonized countries; see, e.g., Subramaniam 2000; or the limitations in removing the racialized inequity in land ownership in Zimbabwe that continued after its independence; see chapter 3), persistence of colonial norms in postcolonial institutional practices (e.g. the "noting" practice in the funding of scientific research in India; see, e.g., Prasad 2005, 2014), racialization of immigration policies in western countries, and so on.

The two examples that I discussed earlier highlight how colonial tropes continue to bear on our imaginaries and practices in transnational contexts. There are numerous examples all around us that exemplify genealogical links to colonialism. Some of these I discuss in this book. It also needs emphasizing that investigation of the genealogies of colonialism to understand the history of the present cannot be undertaken in a causally reductive way. For example, the reneging of the pledge for British economic support for land reforms in postcolonial Zimbabwe by Tony Blair's Labour government in the United Kingdom on the ground that "[w]e are a new Government from diverse backgrounds without links to former colonial interests" (see chapter 3) is based on a facile argument that causality has to be immediate and direct, which conveniently forgets that the benefits of colonialism didn't simply evaporate at the end of colonialism. Moreover, seemingly anti-colonial actions (e.g., buying of the East India Company by an Indian) may remain imbricated within the colonial discourse. We will need to excavate the entangled histories and socialities of colonialism – "the complex, tenuous and often invisible folding of ideologies, classificatory schemas, institutions, political economies, cultures and so on from different time periods and across geographies, as well as between the structured and emergent practices" (Prasad 2014: 7).

As I have argued elsewhere, science and technology studies (STS) "by the very act of showing the multiplicity, contingency, and context-dependence of scientific knowledge and practice," – i.e., by undermining the universality and cultural transcendence of modern science – "provincialized modern science" (Prasad 2016: 194), situating the individual and his/her/their thought in his/her/their local contexts. Right from the time when it emerged in the 1970s, STS consistently "described the *bricolage* of laboratory shop practice" and undermined "Lévi-Strauss's distinction between the *bricoleur* and the engineer" (Lynch 1993: 150). Lévi-Strauss drew on the French term *bricoleur*, which, as he explained, implied "someone who works with his hands and uses devious means compared to those of a craftsman" to describe how "the savage mind" operates in opposition to that of a scientist or an engineer (Lévi-Strauss 1966: 16–17). The mythopoetic construction of the figure of the engineer (who analogically represented the western/modern "other" of the primitive or the "savage mind"), which Jacques Derrida showed through a deconstructive reading of Lévi-Strauss's texts (Derrida 1978), STS showed empirically through a focus on laboratory practices (M. Lynch 1993).

However, even though *bricoleur* and *bricolage* became the defining elements of STS's contribution to the understanding of scientific practice and knowledge production, the colonial genealogy of these terms was all but forgotten. For example, Lévi-Strauss's goal in *The Savage Mind*, in which he presented the distinction between the *bricoleur* and the scientist/engineer, was, in fact, to critique the then prevailing Eurocentric and colonial discourse within which the "difference between the universe of the primitives and our own . . . [was thought to contain] the explanation of their ["primitives"] mental and technological inferiority" (Lévi-Strauss 1966: 267). Lévi-Strauss concluded that the "savage mind is logical in the same sense and the same fashion as ours" (Lévi-Strauss 1966: 268).[28] Nevertheless, his work reinforced logocentrism/Eurocentrism because, as Derrida argued, it describes "what is peculiar to the structural organization . . . by not taking into account . . . its past conditions . . . by putting history between brackets" (Derrida 1978: 291).[29] More broadly, what Derrida suggests in relation to the nature/culture dualism (engagement with which has been important within STS, see, e.g., Latour 1993) holds for the *bricoleur*/engineer dualism, too: "Once the limit of the nature/culture opposition makes itself felt, one might want to question systematically and rigorously the history of these concepts" (Derrida 1978: 284).

In the field of STS, which has been opposed to any structuralist interpretation of scientific knowledge and practice, the success of the empirical illustration of the breakdown of the *bricoleur* versus the engineer/scientist dualist divide possibly resulted in the omitting of a rigorous engagement with the history of these concepts. Modern science, which was commonly represented, to use Sharon Traweek's famous phrase, as a "culture of no culture" (Traweek 1988) was consistently shown by STS to be contingent upon circumstances and socially and culturally situated. However, we have to be careful because we can miss the complex genealogy of Eurocentrism and colonialism that undergirded (and continues to undergird) the understanding of modern science as a universal knowledge that transcended any particular culture.

Modern science was (and continues to be) shown simultaneously as a product of western culture and yet a culture of no culture. The origin of modern science has been argued to be in Western Europe, and yet modern science was not to be associated with any particular culture because it embodied universal values, norms, and practices. It is this, so to say, "god trick" that, for example, allowed Herbert Butterfield to simultaneously make the following two claims: "The scientific revolution we must regard ... as the creative product of the West" (Butterfield 1957: 191), and also that "[i]t was a civilization [ushered in by the birth of modern science] that could cut itself from Graeco-Roman heritage in general, away from Christianity itself – only too confident in its power to exist independent of anything of the kind" (Butterfield 1957: 202). This "god trick" became the paradigm for the diffusion and the dependency theories (see, e.g., Basalla 1967; Rostow 1990 [1960]).

The issue for me is not whether there is no such thing as the Scientific Revolution (i.e., as a revolution that marked a radical rupture; see, e.g., Shapin 1996) or whether the diffusion theory of modern science is too West-centered and linear and, as a result, does not take into account the contributions of the non-western societies (Chambers 1987; Krishna 1992; MacLeod 1987; Raina 1996). Rather, as I show in chapter 2, my concern is how these two contradictory propositions – the situated birth of modern science in the West and yet it being re-presented as a culture of no culture – undergirded a mutation of Euro-/West centrism that represents a *historicism without history*. The non-West was relegated to the "waiting room of history" (Chakrabarty 2000) and defined through "lack" (e.g., of scientific culture, scientific temper, scientific institutions, etc.) and yet the

newly independent postcolonial societies could seek
to be a part of the "commonweal" of modern science
because it was argued to transcend any particular cul-
ture/society and was seen/shown as universal. The result
of such a framing of science has been an endless play
of lack in describing the diffusion of science to non-
western societies, wherein what gets erased is that the
success of Euro-/West-centric historicism of modern
science rests on this (assumed) possibility of diffusion:
"the premise of diffusion of universal, western science
to non-western cultures is what gives credence to the
claim of Western Europe as the exclusive point of origin
of modern science" (Prasad 2019: 1078) and as such the
cradle of universal values and practices. The concept of
the Scientific Revolution, as I show in chapter 2, made
the inherent contradiction in the simultaneous claim
of the universality of modern science and the cultural/
historical particularity of its origin in Western Europe
disappear.

Science Studies Meets Colonialism, however, does
not situate the concern with the colonial genealogy of
science in the non-West or in the distant past. Rather,
it focuses on how the past animates the present, i.e.,
the history of the present. This book, thus, starts with
a study of COVID-19 misinformation and conspiracies that resulted in widespread calls for action against
anti-science responses and "movements" that created a
"pandemic of misinformation." Chapter 1 shows how
an idealized understanding of science was (and continues to be) used to spread COVID misinformation and
conspiracies. It builds on a long-standing concern within
science and technology studies (STS) in the public understanding of science that has consistently highlighted, to
draw on Brian Wynne, that "scientific information is
never, and can never be, a purely intellectual process,
about reception of knowledge per se" (Wynne 1992:
281). In fact, as Sheila Jasanoff argued more than three

decades ago, even in the regulatory engagements with science and technical expertise, the "cognitive authority of science" is threatened (Jasanoff 1987) and as such we need to keep our focus on the coproduction of science and society (Jasanoff 2005, 2011).

The book traces a genealogy of an idealized understanding of science in chapter 2 and analyzes the erasures and contradictions within postcolonial science studies in the last chapter. The book builds on the extensive engagement with colonialism and Euro-/West centrism that started within the field of STS in the 1990s when Sandra Harding called for STS engagement with Euro-/West centrism and colonialism (Harding 1994, 1998). Postcolonial science studies became an important subarea of STS after Warwick Anderson edited a special issue for *Social Studies of Science* titled "Postcolonial Technoscience" (Anderson 2002).[30]

Nonetheless, there has been, at an important level, a nonrecognition of some aspects of the colonial genealogy of modern science that, as I show in chapter 3, results in strange, almost colonial-style, appropriations of postcolonial analytics. I focus on the writings of particular STS scholars, including those of Sandra Harding and Warwick Anderson. My aim is not to discredit the contributions of particular authors. Indeed, I and a large number of STS and non-STS scholars have been heavily intellectually indebted to the work of the authors I critique. I critically engage with the works of Harding and Anderson in particular in order to highlight that we, in the field of STS, need "to see Others not as ontologically given but as historically constituted" (Said 1989: 225). My broader concern in the book, although limited in its scope to the field of STS, is similar to that raised by Ann Stoler in *Imperial Debris*, "whether postcolonial [science] studies has too readily assumed knowledge of the multiple forms in which colonial pasts bear on the present and has been too quick to assert what is actually postcolonial in current situations" (Stoler 2013: 5).

The title of this book, *Science Studies Meets Colonialism*, is aimed, in part, at reiterating STS engagement with colonialism, in particular to understand the history of the present and in part to critically interrogate STS's, so to say, encounters with colonialism. With regard to the latter, although the goal of the book is limited, I hope it can usher in a broader debate that has occurred in some other disciplines. In that regard, there are two issues on my wish list for such a debate (if and when it occurs). First, analogically relates to a claim that Derrida made in relation to ethnology – that we should not forget that the time when ethnology was born was also "the moment when European culture had been *dislocated* . . . had been forced to stop considering itself as a culture of reference" (Derrida 1978: 282). STS emerged in the 1970s and the early 1980s when Europe/the West was being challenged in a variety of ways – intellectually, in relation to the West's engagement with the non-western "other" through the works of, for example, Edward Said, through feminist critiques of western male-centered theories and concepts, and by post-structuralists, cultural studies scholars, and race/racism theorists. To add to these intellectual challenges were the political challenges of student protests, globally, against the existing political systems and the economic situation. In this mix was also the way modern science was imagined to which STS scholarship carefully attended. However, I think we failed to engage with the broader context within which STS emerged, and it needs to be explored whether that elision results in colonial reinscriptions that we see even when some influential STS scholars engage with the non-West or with colonialism (as I show in chapter 3). The other issue on my wish list, which is related to the first one, is whether, in line with the debate that was initiated by, for example, Talal Asad and his collaborators that was published in the edited volume, *Anthropology and the Colonial Encounter* (Asad 1973), we need to much more carefully engage with our own field's genealogical links with colonialism (even though STS was born in a postcolonial era). The book does not offer answers to these questions. I simply seek to pose these questions to the best of my ability in this book.

The term "science studies" is not aimed at signifying a need for etymologically situating the field. It is simply shorthand because, as the editors rightly reminded me, science and technology studies is a mouthful. Moreover, the book is not a critique of STS, in the sense that it is not aimed at simply highlighting problems with the field. Rather, its goal is a further opening of the field with regard to genealogies of colonialism in science, technology, and medicine that would allow us to even more critically explore erasures of certain agents and their "voices." This task is not easy or straightforward. As Gyan Prakash, elaborating Gayatri Chakravorty Spivak's (1988) argument, explained: "The project of retrieval begins at the point of the subaltern's erasure; its very possibility is also a sign of its impossibility, and represents the intervention of the historian-critic whose discourse must be interrogated persistently and whose appropriation of the other should be guarded against vigilantly" (Prakash 1992: 12). The concern should not be who escapes these constraints, but rather the "critical rigor with which . . . [the] relation" to the dominant discourse is interrogated (Derrida 1978: 279).

In this regard, we can learn from the feminist studies which have had a long tradition of such critical and productive engagements that continues to this day; for example, critiques of liberal feminism for being complicit in the colonial/imperial project (Burton 1994; Mohanty 1988; Narayan 1997; Stoler 2010b), or for excluding/erasing voices of women of color (Collins

1996; Davis 1983; Nash 2019), or to "trouble" the very categories of gender/sex (Butler 1990; Haraway 1991). *Science Studies Meets Colonialism*, in a similar vein, interrogates the exclusions persisting in colonial and Orientalist discourses in order to make STS more inclusive.

The book, apart from the introduction and the conclusion, has three long chapters. The first chapter focuses on the misinformation and conspiracies that spread all across the world as the COVID-19 pandemic started to spread. The chapter excavates genealogies of colonialism in the articulation of misinformation and conspiracies through idealized imaginaries of science, in their interpretations by different social groups that reveal colonial links, and in the concern with the origin of COVID-19 that is presented through colonial tropes. There seems to be a general consensus that anti-science attitudes has been propelling misinformation and conspiracies. Moreover, belief in misinformation and conspiracies have often been explained through political attitudes of individuals and attributed to the anxieties caused by the pandemic. Although these explanations are not incorrect, this chapter shows that even when the information is wrong (it is misinformation), the interpretation often reflects genealogical links to colonial history, for example, the history of white-settler colonialism in the United States. This chapter also investigates how, contrary to popular belief, misinformation and conspiracies relied on the credibility of science and scientists. The chapter ends with an exploration of how the concern in the United States about the origin of COVID-19 draws on colonial tropes and is tied to maintaining US-/West-centered geopolitical dominance.

Chapter 2 explores the emergence of two colonial/postcolonial global histories of science in Britain in around the mid-twentieth century, which also marked the start of the end of European colonialism. On the one hand, Joseph Needham's civilizational approach, while presenting a comprehensive history of sciences in China, also constitutes different societies/countries as closed vessels, thereby erasing the connected histories that impacted and radically transformed both Europe and the rest of the world. Needham's famous river metaphor – several streams of science that emerged in different parts of the world joining to make modern science – presented, as the chapter shows, a syncretic and yet Eurocentric historical imaginary of modern science. On the other hand, Herbert Butterfield not only argued that the birth of modern science occurred in Europe during the Scientific Revolution, he also relegated other parts of the world to the role of, at most, messengers of knowledge and modern science, seen/shown by Butterfield as an exclusively European achievement. These two approaches that coincided with the birth of the history of science department at Cambridge University were, as I show in the chapter, complementary and they gave birth to (or at least consolidated) a Eurocentric historicism without history. This modern science-centered Eurocentric historicism became the new template for postcolonial times (for both the erstwhile colonizer and the colonized), and it not only undergirded dependency and modernization theories but also defined the national imaginaries and policies of newly independent countries such as India. Indeed, as this chapter highlights, the genealogy of an idealized imaginary of science that, for example, became a tool to spread COVID-19 misinformation and conspiracies has to be situated in this historiography of the "miraculous" birth of universal, modern science.

The third and last chapter analyzes the genealogies of colonialism in the works of STS, including those by two of the most prominent postcolonial science studies scholars. It investigates how the study of the Zimbabwe bush pump (Laet and Mol 2000), a widely cited STS work, forgets colonial genealogies and as a result erases the voices and histories of the Zimbabweans. In contrast, as the chapter shows, an exploration of the genealogies of colonialism brings to the fore a much more vibrant and complex history of the bush pump that also highlights the racialized and colonial politics of water and arable land in Zimbabwe that forms the social-historical context of the bush pump's development and use. The chapter also analyzes the study of Portuguese voyages of the late fifteenth and the sixteenth centuries (Law 1986) and shows that a lack of engagement with colonial genealogies makes the study rearticulate Orientalist tropes and colonial divides and miss the connected histories of the European and the non-European societies. The last section of the chapter focuses on the postcolonial interventions of Sandra Harding and Warwick Anderson, two central figures in the subfield of postcolonial science studies. It explores how their assumption of the erstwhile colonized and non-western "other" as an "ontological given" either makes them link postcolonial knowledge to fixed locations or present an ambivalence that undercuts the very goals that they seek to achieve through their postcolonial interventions.

More broadly, the book excavates genealogies of colonialism in three contexts. The first context relates to the broader concern that has been called the post-truth era, which is also commonly characterized as anti-science. Prominent STS scholars such as Donna Haraway and Bruno Latour have been directly asked about the role of STS in promoting post-truth (see chapter 1). Recently, Mike Lynch, responding to such concerns, argued that the problem may not be "anti-science per se, but the collapse of more nuanced debate into over-generalized 'scientific' claims in the public airing of disagreements" (Lynch 2020: 55). The title of Lynch's essay, "We Have Never Been Anti-Science," makes the goal of STS clear. Nevertheless, Lynch suggests that the "symmetry and relativism in STS were circumscribed as a part of an effort to approach diverse forms of knowledge" (Lynch 2020: 55). As I have shown elsewhere, there may be a missing link here – the idealized imaginary of science that has a colonial genealogy – and the issue is not with the symmetry principle of STS (Prasad 2022).

The second context is a set of intellectual interventions that took place around the mid-twentieth century in the shadow of the World War II. I focus on these interventions because with World War II the colonial context itself changed. During this period, the history of science department was also established at Cambridge University, and in a significant way it became the site for a genealogical mutation of Eurocentric historicism that served as a template for postcolonial times. The third and the last context on which I focus represents, so to say, coming home and is aimed at excavating genealogies of colonialism in the field of STS itself. *Science Studies Meets Colonialism*, with its three chapters and the introduction and the conclusion, by critically interrogating three colonial/postcolonial contexts, seeks to initiate further debate on engagements with colonialism within the field of STS. The genealogies of colonialism in the present, as I have shown in the introduction and the rest of the book, no longer reflect straightforward hierarchies and simple dualist divides; they are complexly and ambivalently articulated and they disrupt as well as reinforce old colonial hierarchies and tropes.

1

COVID-19, Science versus Anti-Science, and the Colonial Present

I think facts died a long time ago, and it's taken people quite a while to notice.
> Mary Poovey, author of *A History of the Modern Fact*, in an interview with NPR, April 29, 2012[1]

Public trust in science and evidence is essential for overcoming COVID-19.
> Dr Tedros Adhanom Ghebreyesus, Director-General of the World Health Organization, 2021[2]

In March 2021, when spikes in COVID-19 infections and deaths in different parts of the world were continuing to disrupt any possibility of the pandemic ending soon, an article in the *Scientific American* forcefully reiterated another "escalating" concern – the "antiscience movement."[3] The author, "a vaccine scientist and a parent of an adult daughter with autism," claimed that "antiscience is causing mass deaths ... in this COVID-19 pandemic" and characterized anti-science "as a dominant and highly lethal force, and one that threatens global security, as much as do terrorism and nuclear proliferation."[4] The *Scientific American* article's claim may seem a little over the top. However, similar concerns have been raised very frequently during the pandemic. *PLOS Biology*, for example, published an article whose title starts with the phrase "Anti-science kills."[5] In fact, in June, 2020, when the pandemic was ravaging the United States, Dr Anthony Fauci, who has been the Director of the National Institute of Allergy and Infectious Diseases (NIAID) from 1984 and advised eight US presidents, had warned that there was an "alarmingly large percentage of people" with "anti-science . . . feeling."[6]

Concerns about anti-science attitudes impacting COVID-19 responses has not been unique to the United States. In a news report in *Nature*, Brazilian scientists bemoaned "that the [Brazilian] government's failure to follow science-based guidance in responding to the pandemic has made the crisis much worse" and blamed "President Jair Bolsonaro's administration" for having "publicly undermined science."[7] On the other side of the globe, in India, news headlines such as "The Rise of Anti-Science as Covid-19 Cases Exploding" or "The Rise of India's 'Covid Quack'" have been common.[8] An editorial in *Nature* called on India and Brazil to highlight "the human cost of sidelining science."[9]

The concern with anti-science claims and activities, and more broadly with the spread of misinformation, during the COVID-19 pandemic are not unfounded. In February 2020, when COVID-19 infection cases were relatively low (with total confirmed cases only around 15,000) and largely limited to China, the World Health Organization (WHO) had already warned against the accompanying "massive 'infodemic' – an overabundance of information – some accurate some not," adding "that makes it hard for people to find trustworthy sources and reliable guidance when they need it."[10] The situation became progressively worse. COVID-19 has been widely argued to have also spawned a "pandemic of misinformation" (Islam et al. 2020; Li, Bailey, Huynh, and Chan 2020; Moran 2020). There have been widespread protests against mask wearing and vaccination, presentation of different quack therapies, and considerable misinformation and conspiracies have gone viral. The growing global concern with misinformation and conspiracies that are commonly argued to have been fueled by anti-science attitudes has further accentuated the existing disquiet in relation to post-truth.

The Oxford English Dictionary defines post-truth as "relating to or denoting circumstances in which objective facts are less influential in shaping public opinion than appeals to emotion and personal belief."[11] Post-truth has at least three decades of history, as the *Oxford English Dictionary* noted while christening it as the word of the year in 2016 in light of the pro-Brexit vote in the United Kingdom. It "seems to have been first used in this meaning in a 1992 essay by the late Serbian-American playwright Steve Tesich in *The Nation* magazine."[12] Initial concerns with the emerging era of post-truth were largely in relation to politics and news. Even when societal positions against the theory of evolution or climate science were discussed, often the argument was that such positions were not about science. For example, an article published in *Nature*, while affirming that "science teaching worldwide treats evolution as routine," argued that the "controversy [over the theory of evolution] is not really about science but about religion and politics" (Lerner 2000: 287). I am not suggesting that beliefs in creationism or opposition to climate change research were not characterized as anti-science. Rather, COVID-19 seems to have further coalesced and sharpened the concerns in relation to "anti-science" attitudes and activities.[13]

At this juncture, when anti-science and post-truth concerns are so widespread, is making a case for colonial entanglements of science and as such for post/decolonial science studies, as I am doing in this book, worthwhile or even possible? The "science warriors" of the 1990s had accused science and technology studies (STS), and more broadly what they called the "academic left," of "aiding right-wing efforts to obfuscate well established scientific evidence" (Lynch 2020: 52). The "science wars," fortunately, passed.[14] Nevertheless, STS efforts in showing the situatedness of scientific knowledge and practice continued to be questioned and resurfaced again as concerns with post-truth gained momentum. For example, in 2019, a year before COVID-19 and with it the pandemic of misinformation and conspiracies that hit the world, Donna Haraway was asked during an interview with the *Guardian*: "We are often told we are living in a time of 'post-truth.' Some critics have blamed philosophers like yourself for creating the environment of 'relativism' in which 'post-truth' flourishes. How do you respond to that?" Haraway replied: "Our view was *never* that truth is just a question of which perspective you see it from" (emphasis in the original).[15]

More recently, Michael Lynch, another of the founders of the field of STS, responding to concerns about the role of STS in relation to post-truth, argued that "symmetry [in relation to true and false claims] and relativism in STS were circumscribed" (Lynch 2020: 53). Lynch added that the STS principle of "symmetry does not purport to be a 'post-truth' that applies in the same way in all the cases" (2020: 54).[16] Indeed, both Haraway and Lynch are right in presenting the nuanced position of STS in relation to scientific knowledge and practice. Nonetheless, these responses also highlight the impact of persisting concerns in relation to the field of STS itself. Post/decolonial science studies adds another layer to the arguments about situatedness and construction of scientific knowledge – the imbrication of science(s) within colonial discourses and practices and its continued impact in postcolonial contexts. Postcolonial approaches, as Warwick Anderson pointed out, have not gained very significant traction within STS.[17] Even when mainstream STS scholarship has engaged with postcolonial issues, often it has either bypassed postcolonial analytics and concerns or, at times, even replicated colonial tropes, as I show in chapter 3.

In short, engagement with colonial entanglements of science is particularly fraught at this time of widespread concerns with anti-science attitudes and activities. Any call to critically investigate socio-historical entanglements of scientific knowledge and practices can be seen as providing fodder to anti-science beliefs. However, as STS scholars have argued for more than two decades, "'facts' cannot stand apart from wider social, economic, and moral questions even if rhetorically they are often put forward as if this were the case" (Irwin and Wynne 1996: 3). As this book, building on a long line of STS work on public understanding of science, argues, it is not only important but also urgent to engage with post/decolonial concerns in relation to science. COVID-19 misinformation and conspiracies have commonly utilized *idealized* constructions of science that have been entangled with colonial imaginaries and practices (see chapter 2 for a detailed discussion on the emergence of an idealized science). Moreover, misinformation and conspiracies, even though they are false, are often interpreted through genealogies of colonial experiences and draw on colonial tropes (Prasad 2022).

Ironically, responses towards anti-science claims are often no different in this regard. The result is a dualist construction of self/other (through science/anti-science) that inhibits proper understanding of what is going on and thereby limits effective action against misinformation and conspiracies. In the following, I critically investigate three aspects of COVID-19 misinformation and conspiracies, namely, deployment of an idealized imaginary of science to frame misinformation and conspiracies, colonial genealogies in the interpretation of misinformation, and use of colonial tropes in the claims about the origin of coronavirus (SARS-CoV-2 virus), which caused the COVID-19 pandemic.

Science in the web of misinformation and conspiracies

South Korea is among the very few countries that effectively managed COVID-19 infections and deaths. By mid-June 2021, when globally there were 176 million reported cases of infection and 3.81 million deaths, South Korea had slightly less than 150,000 cases of infection and fewer than 2,000 deaths. Nevertheless, as South Korea raced to gets its citizens vaccinated, misinformation giving reasons for not getting vaccinated appeared on the "South Korean blogging platform Naver Blog on May 29, 2021" and quickly started to spread on social media.[18] *Agence France-Presse* (AFP), the oldest international news agency that was founded in 1835 and has its headquarters in Paris, France, reported the misinformation and published its translation in English.

The COVID misinformation post, according to AFP, read: "I met with a prominent professor of biotechnology. He's currently working at a university in Boston, USA. We talked about the current Covid-19 situation. Since he is afraid of being called a conspiracy theorist, he didn't want to identify himself. But I'm sharing what I heard because it might be helpful to you."

The post lists "eight reasons not to get vaccinated":

1 mRNA technology lacks a track record of studies;
2 there is no scientific evidence that COVID-19 vaccines can protect people;
3 the virus's spike protein sheds and reproduces itself;
4 the body will become a factory to produce spike protein;
5 COVID-19 vaccines were not tested on animals;

6 vaccine makers did not reveal the ingredients of their products;
7 doctors are afraid of reporting adverse events in case of retaliation from drug companies; and
8 PCR tests are not scientifically reliable.[19]

AFP has rebutted each of the above-listed reasons for not getting vaccinated and shown them to be false. There is no reason to dispute AFP's rebuttal of the misinformation. However, in order to better understand COVID misinformation and its spread, we need to investigate how the "veracity of a representation . . . becomes hostage to its circulation and mobilization within and by particular social groups" (Prasad 2022: 90).[20] It is, thus, useful to analyze the tropes that are used to spread misinformation and analyze their discursive emplotment.

The above-quoted COVID-19 misinformation blog post, right at the outset, presents a "prominent professor of biotechnology" as the source. Now that we know that the blog was spreading misinformation, one can safely assume that the cited source is false too. It is likely that the blogger had an ulterior motive in presenting a scientist as the source. Nonetheless, the deployment of a "prominent professor of biotechnology" as the source highlights that the credibility of a scientist can serve as a useful trope to instill confidence in the readers. And the phrase "afraid of being called a conspiracy theorist" plays into the dualist discourse of science versus anti-science to hide the identity of the suggested source. Moreover, the fact that the "prominent biotechnologist" is presented as working in a university in Boston, United States, reflects not only the transnational scope of misinformation, but also the deployment of existing transnational hierarchy in the scientific/academic field – the Boston area is home to several top-ranked universities, including MIT and Harvard.

Utilization of the credibility of science and scientists is not unique to this particular blog. *Plandemic*, a 26-minute COVID misinformation and conspiracy video that was posted online on several social media platforms on May 4, 2020 and soon went viral, deployed a similar discursive strategy. This video, which is an interview with a discredited virologist, Dr Judy Mikovits, starts with the claim: "Dr Judy Mikovits *has been called one of the most accomplished scientists of her generation. Her 1991 doctoral thesis revolutionized the treatment of HIV AIDS. At the height of her career, Dr Mikovits published a blockbuster article in the journal Science*" (emphasis added).[21] *Plandemic*, in just over a week after its release, was "viewed eight million times on YouTube, Facebook, Twitter and Instagram" and, on Facebook, it was "liked, commented on or shared nearly 2.5 million times."[22] The video was removed from all major social media platforms. However, even a year after its removal, *Plandemic* continues to be available, for example, through PATRIO.Tv.[23] The video is also often referenced or clips from it are shown in other conspiracy claims. The viral spread of the *Plandemic* video is reflective of a broader issue in relation to misinformation and conspiracies. Virality, which is being presented as a conceptual tool to understand present-day transformations (see, e.g., Peters, McLaren, and Jandric 2020), along with reflecting "the speed with which information travels," can "create or at least add credibility" and "indicates the mutative potential of the media" (Prasad 2022: 92).

Judy Mikovits is also co-author of the book, *Plague of Corruption: Restoring Faith in the Promise of Science* (Mikovits and Heckenlively 2020). The book, which was briefly discussed in the *Plandemic* video, continues to be a best seller. The short bio of Mikovits on Amazon states: "Judy Mikovits, PhD, spent twenty years at the National Cancer Institute, working with Dr Frank Ruscetti, one of the founding fathers of human retrovirology, and has coauthored more than forty scientific papers. She co-founded and directed the first neuroimmune disease institute using a systems biology approach in 2006."[24] Far from presenting her as an anti-science warrior, Mikovits's bio prominently displays her scientific credentials. *Plague of Corruption*'s co-author, Kent Heckenlively, "who bills himself as 'world's #1 anti-vaxxer'" and was "denied a visa to enter Australia for a speaking tour" in 2017 because of his anti-vaccination position,[25] on his author page on Amazon states that, eventually, after his stint as a lawyer, he was "drawn to his original love of science and became a science teacher." His author bio ends with the claim, "Now I get to teach science during the day and write about it at night."[26]

Deployment of the credibility of science and scientists to spread misinformation and conspiracies is, in fact, quite common. For example, Reverend Danny Jones, a senior pastor of the Northlake Baptist Church in Gainesville, Georgia, who does not shy away from calling his claims conspiracies, in a sermon on Bill Gates's role in implanting a microchip through the vaccine in order to create a new world order, states:

> On December 23, 2019 the prestigious *Scientific American* magazine reported that the Bill and Melinda Gates Foundation and MIT [the Massachusetts Institute of Technology] had developed a biometric tattoo where a nanochip can be injected into your forearm at the same time you are being vaccinated, therefore, your arm can be scanned to reveal your identity, your vaccinations, maybe even your medical records. The biometric tattoo is a part of a bigger plan called ID 2020 that was announced this January [2020] at the World Economic Forum in Davos, again sponsored by the Bill Gates Foundation, the Rockefeller Foundation, and several other billionaire organizations.[27]

This sermon has received a lot of attention and the video has been posted on various online sites. The site from which I accessed the video had more than 1.5 million views by mid-June 2021. In the sermon, the credibility of science – expressed by invoking "the prestigious *Scientific American*" – is used to spread a conspiracy. Interestingly, readers and viewers of conspiracies and misinformation are often no different in their reliance on the credibility of science to affirm their faith in particular conspiracies and misinformation. One of the reviewers of Mikovits and Heckenlively's book, *Plague of Corruption* (2020), titled his/her/their review "Before we can restore faith in science we have to know how it was lost." Another reviewer wrote that "If you are a critical thinker and you would like to see through the eyes of an American scientist . . . you will love this book."[28]

Invocations of the credibility of science and scientists are, however, complexly entangled with social interests. For example, the viewers who commented on the video of Rev. Danny Jones's sermon that was posted by the Citizen Media News often criticized the elites in power and affirmed their faith in Jesus and Christianity (some even suggesting that liberals were attacking Christian practices). Their criticisms of the political elite are often followed with calls such as "Wake up, Christians," "Trust only in the Lord," or "Get up close and personal with JESUS CHRIST."[29] Religious, specifically Christian, idiom was also used by "a former Capitol Hill staffer, political analyst on public radio and Fox News affiliates" to praise Rev. Danny Jones's sermon and thereby affirm her anti-vaccination position, but she used it to call for a pro-life position. The blogger writes: "Here is a question for Christians who claim to be pro-life: why are they getting vaccines that have aborted baby tissues in them? Where is the outrage to end the use of baby parts by pharmaceutical companies?"[30] On the other hand, in a post in the *Northside Sun*, which had nearly 280,000 reads, the writer, while concurring with Rev. Jones's sermon, presents his critique of the business elite's control of the global order with a secular twist. He writes, "I think the pastor's 'one world conspiracy theme orchestrated by Davos elites' has some elements of truth. That's probably why it went viral." The author of this post, while railing against "mega rich liberals (and conservatives)," goes on to claim, "I think the Covid-19 epidemic may have originated from 'gain of function' research (to make viruses more deadly) in a lab in Wuhan funded in part by the US."[31]

An overarching characterization of misinformation and conspiracies as anti-science is thus not only intellectually lazy, it also inhibits a proper understanding of how misinformation and conspiracies spread through entanglements of various social interests. Theda Skocpol, drawing on her study of opposition to climate science, suggested in a recent interview that "[i]t isn't that religious conservatives were unaware of science or rejected scientific findings . . . It's that they resent the use of experts as political authorities. And I think that is exactly what we see here [in the case of COVID-19 misinformation and conspiracies]."[32] Several studies, particularly those conducted in European countries and the United States, have argued that vaccine hesitancy (even before COVID-19), for example, was "driven by . . . profound distrust in elites and experts" (Kennedy 2019: 512). COVID-19 misinformation and conspiracies in the United States against Dr Anthony Fauci seem to exemplify the above-stated concern with experts as political authorities. However, misinformation and conspiracies often posit and support their own experts as political authorities or vice versa (i.e., political authorities as experts, as happened in relation to former President Donald Trump). In fact, the "anti-vaccination movement has had a long tradition of promoting the words of 'experts' who support their narrative."[33]

Moreover, trust in scientists as experts has remained high and crosses political divides. A PEW survey in 2019, for example, found that "86% of Americans say they have at least 'a fair amount' of confidence in scientists to act in the public interest," and the variation between those identifying as Republicans and Democrats was 9% (82% for Republicans and 91% for Democrats).[34] Even during the pandemic, while making a call to "rebuild trust in science," PEW admitted, "Fortunately, we start from a solid foundation. *Seventy-three percent of adults in the United States – like majorities around the world* – agree that science and technology make our lives better, and they trust scientists and researchers to make important discoveries that help solve problems" (emphasis added).[35] In fact, in May 2020, when not only the pandemic but also misinformation and conspiracies were raging in the United States, a survey conducted by the COVID-19 Consortium for Understanding Public Policy Preferences Across States found "that both Democrats and Republicans trust scientists and doctors to handle the COVID-19 pandemic more than any other individual or institution, including Congress, President Trump, state government and even the Centers for Disease Control and Prevention."[36]

A scientific expert can, nonetheless, become the target of partisan politics via conspiracies, as has been evident in the persistent claims of several Republican senators and Congress members in relation to Dr Fauci.[37] Interestingly, even in such partisan attacks, the credibility of science has been used as a foil not only to discredit but also to reiterate COVID-19 misinformation. Dr Fauci, responding to various conspiracy and misinformation about him, for example, stated that the "attacks on me are, quite frankly, attacks on science."[38] On the other hand, US Senator Marsha Blackburn, who is one of the proponents of a conspiracy that claims Dr Fauci's role in the origin of the coronavirus, responded, "I am not attacking science. What I am doing is standing up for science." She goes on to state that "Dr Fauci should have learned in science class that you need evidence to support a claim. Fauci was once again evasive and dishonest."[39]

The issue here, as I argue throughout this chapter, is not whether Senator Blackburn is making a false claim about science or if Dr Fauci is overstating his claim about science (rather than highlighting the partisan nature of attacks against him). These invocations of science show that the credibility of science – not just as a signifier but also as the *point de capiton* – remains a potent and indisputable explanatory mechanism for all sides. An idealized imaginary of science, discursively constituted through Euro-/West-centric and colonial entanglements (see chapter 2), allows contradictory and overdetermined invocations of science in all kinds of claims, including conspiracies. In my research on COVID-19 misinformation and conspiracies, I have not found a single instance where the proponent of misinformation and/or conspiracies calls himself/herself/themselves anti-science, even though the claims are evidently against established "scientific facts." That is the reason, I believe, we need to *decolonize* science and abandon its idealized imaginary.

Conspiracies and misinformation often do not reflect a singular, cohesive stand. QAnon, for example, has received a lot of attention for its conspiracy theories.[40] A study of QAnon's conspiracy beliefs, conducted by political scientist Brian Schaffner for the Institute for Strategic Dialogue, found that the "average respondent who viewed QAnon favorably had heard less than half of the four QAnon conspiracies" about which they were asked, "and only believed one of the four." Moreover, "16% of those who did not rate QAnon favorably recognized and believed at least one of QAnon's conspiracy claims."[41] Misinformation and conspiracies, as I show in the next section, reflect complex hegemonic alignments that bring together different social interests and experiences whose genealogies, in some cases, could be traced to colonialism.

Specifically, I show how the discursive emplotment of COVID-19 misinformation and conspiracies reflects genealogies of the colonial present. I situate my analysis largely in the context of the United States in order to show that the colonial discourse continues to impact not only the erstwhile colonized countries in the non-West, but also the dominant center of the West at present. As I argued in the Introduction, I draw on Michel Foucault's concept of the history of the present to investigate the continuing role of colonial imaginaries and practices. Foucault uses the phrase "history of the present" to describe the "ontology of ourselves, of the present reality" (Foucault 2011: 21).[42] My goal, similar to that of Foucault, is to genealogically "trace the erratic and discontinuous process whereby the past became the present." To clarify again, my analysis is not aimed at connecting "the present-day phenomenon to its origins, as if one were showing … a building solidly rooted in the past and confidently projected into the future" (Garland 2014: 372).

Tracing history of the present through misinformation and conspiracies

The Conspiracy Theorist In Me doesn't believe this [news of then President Trump getting COVID]. Trump would have to pass away from Covid (GOD forbid, I don't wish death on anyone) in order for me to believe it. I have 3 thoughts on this (1) I feel this is just his way to change the headlines in regards to not condemning white supremacy. (2) This gives him a reason to not campaign for 14 days, stay out of the press keep the bad headlines down before the election. (3) All a stunt to make one of them Value Menu Vaccines they rushing out look safe. He will take it on national TV, all of sudden be well and give the illusion to the American public that the vaccine is safe, he will look like a hero because he's one of the first to try it, millions of people line up to get it and BOOM microchip implants for all of ya'll!!!!

Charlamagne Tha God, radio show host, October 2, 2020.[43]

The above-quoted claim was not the first time that Charlamagne Tha God, one of the hosts of the radio talk show *The Breakfast Club*, had expressed concerns about COVID-19 vaccines through misinformation and conspiracy.[44] On August 11, 2020, tagging a CNN breaking news story stating that "Vladmir Putin says Russia has approved a coronavirus vaccine, but there are fears that essential corners may have been cut in its development," he had tweeted, "If you take this vaccine on November 3, 2020 you will wake up paralyzed."[45] Charlamagne Tha God has been criticized for spreading misinformation, and the conspiracy theory about implanting microchips via vaccine has been shown to be false.[46]

Charlamagne Tha God is one of a significantly large percentage of people, not just in the United States but in other countries as well, who believe in and/or have spread COVID-19 misinformation and conspiracies about vaccines. A YouGov-Cambridge Globalism "survey of about 26,000 people in 25 countries designed in collaboration with the *Guardian* [had] found widespread and concerning scepticism about vaccine safety": "Across 19 different countries 20% or more of respondents said they gave at least some credibility to the view that 'the truth about the harmful effects of vaccines is being deliberately hidden from the public,' including 57% of South Africans, 48% of Turks, 38% of French people, 33% of Americans, 31% of Germans and 26% of Swedes."[47]

Such widespread concerns with COVID-19 vaccines that are often combined with misinformation and conspiracies have further reinforced calls to combat anti-science beliefs. Headlines such as "Misinformation Spread by Anti-Science Groups Endangers COVID-19 Vaccination Efforts" have been common.[48] Peter Hotez, professor of pediatrics and molecular virology and microbiology and dean of the National School of Tropical Medicine at Baylor College of Medicine, has been among the most actively engaged in this regard. Along with writing a book, *Preventing the Next Pandemic: Vaccine Diplomacy in a Time of Anti-Science* (Hotez 2021), he has also written opinion pieces and journal articles, and his interviews have been published in, for example, *Nature*.[49] In June 2021, in a *USA Today* article, he warned: "Until as a nation we resolve to counteract the anti-science aggression, we may fail to vaccinate our way out of this epidemic by this summer and soon face the specter of emerging variants of concern."[50]

Concern with opposition to vaccination and, more broadly, with COVID-19 misinformation and conspiracies is understandable.[51] However, let me ask again, what do we mean when we characterize opposition to vaccination as anti-science? Often, as I have been arguing, such characterization of COVID-19 misinformation and conspiracies relies on an imaginary of science that, as I discuss in the next chapter, very much like Edward Said's (1979) description of the Orientalist imaginative geography, intensifies a sense of self (science/scientist/ believers of science) by dramatizing the distance and difference between science and anti-science. Such an imaginary of science, and the resulting overarching characterization of misinformation and conspiracies as anti-science, precludes proper understanding of what is actually going on.

For example, a study comparing COVID-19 misinformation and conspiracies in eight countries showed that "respondents from countries with a media-supportive and consensual political system . . . [Belgium, Switzerland, Canada, England] report some of the lowest conspiracy/misinformation beliefs, although scores for English respondents are markedly higher than for those from other countries in this cluster." The study also found "that conspiracy beliefs are higher among respondents in countries with a polarized political and media environment."[52] Misinformation and conspiracies may not only depend on and vary according to social contexts, as the above-quoted study showed, they may also not result in singular and unidirectional actions. A study based on a survey conducted in early 2020, for example, found "that belief in the risk-acceptance type of conspiracy theory was associated with higher risk perception . . . By contrast, belief in the risk-rejection type of conspiracy theory was associated with lower risk perception." The study went on to argue that "these two links appear to play out differently at different stages of the COVID-19 outbreak, such that the former link was stronger during the early, mild stage of the outbreak, whereas the latter link was stronger during the later, severe stage."[53] That is to say, belief in misinformation and conspiracies may not result in linear, temporally static, and unidirectional actions.

More broadly, misinformation and conspiracies are entangled with various psychological and social concerns, which may get further accentuated as a result of contingencies and uncertainties of the time, as has been reported during the COVID-19 pandemic (see, e.g., Coninck et al. 2021).[54] However, it would be short-sighted to assume that misinformation and conspiracies are simply dependent upon anxieties resulting from uncertainties such as those arising as a result of a pandemic. They have also become commonplace more generally. Political conspiracies and misinformation, for example, are being witnessed in democratic election processes all across the world. I do not believe that a war on misinformation (much like a war on pathogens) is going to work. We will have to learn to live with misinformation and conspiracies, and as such studies of their multilayered and multi-modal articulations are necessary and urgent (Prasad 2022). My concern in this chapter is how misinformation and conspiracies about COVID-19 fold and express alignment of different scientific and social interests. In particular, I trace the colonial genealogy of the present in the articulations of COVID-19 misinformation and conspiracies.

In late February 2020, when COVID-19 infections had started to rapidly grow, the number of infections in African countries remained very low. The news agency *France 24*, while highlighting the remarkably low infections in African countries, stated that the low numbers of infection continued "to puzzle – and worry – experts," considering "their fragile health systems."[55] Interestingly, belying such concerns, nearly one and a half years into the pandemic COVID-19 infections and deaths in African countries remained comparatively low. By July 2021, when the United States had nearly 35 million infections and more than 600,000 deaths, India 31 million infections and more than 400,000 deaths, France nearly 6 million infections and 111,000 deaths, and the United Kingdom more than 5 million infections and 128,000 deaths, the African countries had a total of slightly more than 6 million infections and 155,000 deaths.[56]

The remarkably lower infections in Africa could be a result of "limited testing capacity, poor reporting, younger populations, and differences in climate and humidity" (Binagwaho, Frisch, Ntawukuriyayo, and Hirschhorn 2020: 1; Marbot 2020). It is also a result of effective management of the pandemic. Many "African countries have been successful in containing the initial outbreaks because they are rapidly and effectively implementing evidence-based interventions including promoting and facilitating: handwashing, social distancing, testing, contact tracing, and lockdowns" (Binagwaho et al. 2020: 1).

Lower rates of COVID-19 infections and deaths in Africa, however, also became a source of misinformation. In February 2020, a story spread about a Cameroonian college student in China who was purported to have recovered after getting infected "because of his blood genetic composition which is mainly found in the . . . sub-Saharan Africans." It was further claimed that Chinese doctors had confirmed that this person's resistance to COVID-19 was a result of the genetic composition of the sub-Saharan Africans. In actuality, the student was quarantined for 13 days and recovered only after treatment with various medicines.[57] Nevertheless, the above-mentioned misinformation gained momentum and was even given an anti-colonial twist that sought to overturn the racial hierarchy that undergirded European colonialism:

> Caucasians is [sic] always at war with our black skin because they know our melanin is our defense against all that they throw at us. This proves yet again that the black man is indestructible, our bodies are made of the same substances that make up this Earth because we are owners of this universe, they will never wipe us off, history has already proved that.[58]

The remarkably low COVID-19 infections and deaths in Africa were also used to spread misinformation in the United States. In early March 2020, for example, a tweet claimed: "So none of these coronavirus cases have been black people?" The tweet thereafter stated that it could be because of the immunity of black people.[59] The common theory for such an immunity "stemmed from the belief that melanin, the pigment found in hair, skin and eyes, offered a layer of protection from the virus" (Collins-Dexter 2020: 9). This particular misinformation ran out of steam in the United States when data started showing that African Americans were getting infected and were dying as a result of COVID-19 disproportionately. The Centers for Disease Control and Prevention (CDC), for example, reported that in comparison to the whites, Native Americans had 2.8 times higher cases and 1.4 times higher deaths, African Americans had 2.6 times higher cases and 2.1 times higher deaths, and Hispanic/Latinos had 2.8 times higher cases and 1.1 higher deaths.[60] This pandemic, as has been argued by various scholars, has glaringly highlighted persistent racial disparities in healthcare access in the United States (Chowkwanyun and Reed 2020; Hooper, Napoles, and Perez-Stable 2020; McLaren 2021).

Returning to the misinformation about melanin providing protection against COVID-19 infection, it may seem that empirical reality can serve as an effective tool to undercut the spread of misinformation and conspiracies. However, the relationship of misinformation and conspiracies to empirical reality is not straightforward. If it were, there would not be any misinformation and conspiracies in the first place. Analysis of the discursive framing of misinformation and conspiracies provides a window to their complex entanglements with social and historical realities. For example, how are we to make sense of historical references in COVID-19 misinformation? The tweet that I quoted above, for example, ended poignantly, almost as a plea: "It's the least God can do after slavery."[61] Such folding of history in COVID-19 misinformation and conspiracies has been common.

Susan Reverby, responding to COVID-19 misinformation and conspiracies among the marginalized communities of color, wrote that "[w]hen people of color refer to such historical claims it is often as a way to say that structural racism is real . . . but when it comes to health care I will explain this experience in historical terms because it sounds less crazy" (Reverby 2021: 2). Reverby is not alone in attempting to counter the role of such historical references in COVID-19 misinformation and conspiracies. A significant reason for this concern is because "vaccine hesitancy has been laid primarily at the feet of African American and Latinx communities in the United States" (Reverby 2021: 1). Karen Lincoln, professor of social work at the University of Southern California, put it bluntly: "It's [use of historical traumas to explain vaccine hesitancy] a scapegoat. It's an excuse. If you continue it as a way of explaining why many African Americans are hesitant, it almost absolves you of having to learn more, do more, involve other people – admit that racism actually is a thing today."[62] Similarly, a *New England Journal of Medicine* article emphasized that "attributing distrust primarily to these instances [of historical trauma such as the Tuskegee experiment] ignores the everyday racism that Black communities face" (Bajaj and Stanford 2021: e12).

Everyday racial discrimination in healthcare, as a result of long-standing systemic racism, may arise even as an oversight. For example, Chicago's "health department sought to provide local pharmacies with the vaccines. But that well-intentioned effort did not account for the fact that many of the city's minority communities are 'pharmacy deserts' or are populated not by the big chains but by independent stores that do not have the capacity to vaccinate right now" (Reverby 2021: 2). The racial/ethnic disparity in the impact of COVID-19 exists because race and ethnicity are also markers of a range of conditions that affect infections and mortality – "including socioeconomic status, access to health care, and increased exposure to the virus due to occupation (e.g., frontline, essential, and critical infrastructure workers)."[63]

It is not just ironic that one has to critique invocations of historical traumas suffered by black communities in order to highlight the impact of everyday racism to explain their vaccine hesitancy. The need for such a defense highlights how positing of such blame (for vaccine hesitancy) on the black and brown communities is itself a legacy of the white-settler colonialism in the United States. Such colonial stereotyping, which operates by repetition and enfolds a biopolitical strategy of hierarchical control (Bhabha 1994), belies the empirical realities. For example, according to polls jointly conducted by the University of Texas and the *Texas Tribune*, while 53% of blacks and 34% of whites said "no" when asked "Would you try to get a coronavirus vaccine if widely available at a low cost?" in October 2020, in February 2021 the figure for blacks saying "no" to vaccination dropped dramatically to 29%, while for whites it became 28%.[64] The same polls showed that "61% of white Republicans . . . said they are reluctant to get the vaccine or would refuse it outright."[65] As these polls show, fetishization of anti-vaccine responses as *anti-science* beliefs is not just unhelpful, it also hides what is going on; for example, the discursive entanglements of these responses with racialized imaginaries of a nation.

There have been arguments, based on studies of anti-vaccine misinformation spread on social media, that suggest "vaccine opposition may be coalescing around a common narrative, emphasizing civil rights and freedom from elitist government overreach" (Broniatowski et al. 2020; Kennedy 2019). Nonetheless, even in such coalescing of narratives, variations across social/racial groups are also evident: "vaccine refusers tend to be well educated, white, and more affluent than people who typically experience health disparities" (Dredze, Broniatowski, Smith, and Hilyard 2016: 550). Similarly, in relation to opposition to climate science, a study based on a decade of Gallup surveys (2001–2010), showed that "conservative white males were four times more likely than all other adults to believe that 'the effect of global warming will never happen'" (McCright and Dunlap 2011; Prasad 2022). My aim is certainly not to argue that white people are more likely to have anti-climate science, anti-vaccine or, more broadly, anti-science attitudes. Rather, I want to suggest that the discursive emplotment of misinformation and conspiracies present to us cognitive maps that, among other things, may reflect the social-structural position of the believer/proponent.[66]

Charlamagne Tha God's defense for his invocation of the microchip conspiracy, which he later posted on Facebook, illustrates my claim. His articulation of the microchip conspiracy was folded with a response to the news of the former President Donald Trump having been tested positive for COVID-19. Tagging President Trump's tweet of October 2, 2020 declaring the latter testing positive for COVID-19, Charlamagne Tha God starts the post by stating: "The Conspiracy Theorist in Me doesn't believe this." He then explains that he feels that the news of President Trump testing positive was "just his way to change the headlines in regards to not condemning white supremacy." Thereafter, he added how this is a "stunt to make . . . vaccines they are rushing out look safe," which would eventually portray President Trump as "a hero because he's one of the first to try it." He then ends the post with the claim that the vaccines would implant microchips in millions of people.[67] It is important to note that, in the discursive emplotment of microchip conspiracy, racism and white supremacy are key facets.

Charlamagne Tha God's defense of his anti-vaccine position further illustrates how his concern as an individual and as a black person is tied to the long-racialized history of the United States. In an interview after he had made the misinformation and conspiracy claims, responding to a question on the concern about the COVID-19 vaccine among some black people, he said there was definitely mistrust as a result of "systemic racism in the medical system." Recounting him being asked by the President Biden administration and Governor Cuomo (of New York) to be on task forces to get people vaccinated, he reflected: "I've never seen y'all in a rush to remedy anything in the black community ... but with the vaccine it's like we got to get it to the black people, we got to get it to the black people now." Thereafter, he stated that "we have hundreds ... of years [of experience] why we should not trust the government."[68] In the interview he even admitted that he is not against the vaccine and his children have been vaccinated. The anti-vaccine position of Charlamagne Tha God is a striking example of how we need to be alert to the history of the present, in this case, to the history of white-settler colonialism in the United States.

I have focused on Charlamagne Tha God's articulation of the microchip conspiracy because, although it has received significant media attention, racial differences in the articulation of such conspiracies are often not covered or get hidden in the media glare. Charlamagne Tha God is a well-known African-American public figure and his co-hosted show *The Breakfast Club* has "77 percent ... African-American or Hispanic" audience.[69] As such, advertently or inadvertently, one may slip into repeating the stereotype of putting blame on people of color for vaccine hesitancy. An article, published in June 2021, "Where did the microchip vaccine conspiracy theory come from anyway?," for example, starts with

Charlamagne Tha God's articulation of the microchip conspiracy.[70] This article, to be fair to the reporter, is not blaming Charlamagne Tha God for starting the conspiracy. It traces the origin of the conspiracy to the response of the administrator of a Swedish website "who goes by the name CyphR" and "belongs to a community of biohackers who advocate for human implantable microchips."[71] CyphR, responding to an online chat on Reddit AMA on March 18, 2020 in which Bill Gates "predicted that one day, we would all carry a digital passport for our health records," had written: "This is it! Suddenly, chip implants don't only have an actual, scalable mainstream application, but one that is an urgent medical need." CyphR headlined his/her claim, "Bill Gates will use microchip implants to fight coronavirus."

Adam Fannin, a white Baptist preacher from Florida, while drawing on the above-discussed claim on the Swedish website, titled a video posted on his YouTube channel "Bill Gates – Microchip Vaccine Implants to Fight Coronavirus." This video, which was posted two days after the Swedish website's claim, added *vaccine* to the headline of CyphR. It soon went viral with 1.6 million views. The above-quoted media report also cites a survey conducted in March 2021 that "found that 16 percent of eligible Americans . . . [were] a hardened group of COVID-19 skeptics steeped in conspiracy theories, while another 7 percent are system distrusters," adding that "skeptics . . . [were] disproportionately white and conservative, while the system distrusters are heavily Black and Latinx."[72] These variations across racial/ethnic groups, which are further accentuated when we look at the discursive emplotment of misinformation and conspiracies, cannot be reductively argued to be an unchanging effect of white-settler colonialism in the United States; nonetheless, the colonial genealogy of the present cannot be denied.

At the turn of the last century, Theodore Roosevelt, the 26th president of the United States (1901–1909), in his response to historian-journalist Charles Pearson's *National Life and Character: A Forecast* (Pearson 1893) had, for example, proudly claimed: "*Nineteenth century democracy needs no more complete vindication for its existence than the fact that it has kept for the white race the best portions of the new world's surface, temperate America and Australia*" (emphasis added).[73] Pearson's book was an account and prognosis of "white man under siege" because "the 'Black and Yellow' races . . . were in the ascendant" (Lake 2004: 42). Roosevelt, while recognizing "the 'great effect' of . . . [Pearson's] work . . . in Washington" (Lake 2004: 41) was thus assuring Pearson and other white contemporaries who at the turn of the last millennium had anxieties about the possible loss of white supremacy (Roosevelt characterized Pearson as belonging "to the melancholy or pessimist school"). Roosevelt wrote that "the peopling of the great island-continent with men of the English stock is a thousand-fold more important than the holding [colonizing] of Hindoostan [India] for a few centuries" (as quoted in Anderson 2006: 254).

Patricia Hills Collins's claim that "US national identity may be grounded more in ethnic nationalism than is typically realized" (Collins 1998: 70) is thus also a reminder of the amnesia about the racialized "imagination" of the US nationhood that was explicitly and proudly expressed by the then President Roosevelt and continues to bear on various policies and actions. For example, as Collins shows, "differential population policies developed for different [racial and ethnic] segments of the US population emerge in direct relation to any group's perceived value within the nation-state" (Collins 1998: 76). The genealogy of white-settler colonialism is strikingly evident at present. The pandemic has coincided with vigorous and open assertions of white supremacy (a reaction to which was folded in the conspiracy claim of Charlamagne Tha God) and resistance to them via Black Lives Matter that, after the killing of George Floyd and Breonna Taylor, gained steam not just in the United States but globally. In the context of the pandemic, arguments such as the "first amendment of the US Constitution gives every American the right to question the safety of any vaccination,"[74] or anti-mask protesters arguing that they are doing it for the protection of freedom and liberty that are largely presented by whites, are not simply a coincidence. They represent a genealogy of the colonial present (see Prasad 2022 for a more detailed discussion).

The discursive emplotment of misinformation and conspiracies, which I have analyzed in this section, expands Warwick Anderson's exposition of "cultivation of whiteness" through which he argues that "clinic and laboratory should be added to . . . [the] sites where the nation – any nation – may be imagined" (Anderson 2006: 2).[75] A pandemic, as COVID-19 has shown, is another site where the nation – across the world – may be imagined and the exclusionary and colonial aspects of the "imagined community" (B. Anderson 1991) contested as well as defended. Nation-building, as Eugen Weber showed through his historical account of imposition of the French national ideology, which "was still diffuse and amorphous around the middle of the nineteenth century," was akin to colonialism (Weber 1976: 486). Weber states that apart from the fact that "[g]iven time and skins of the same color, assimilation worked" in the case of France, Frantz "Fanon's account of the colonial experience is an apt description of what happened in the Landes and Corrèze [parts of France]" (E. Weber 1976: 491).

The assimilation within a nation is, however, never complete because of exclusions based on racial/ethnic differences, that of immigrant groups from the erstwhile colonized countries, not forgetting the desire of the post-colonial countries to create "one nation, one people" in the projected image of the western nations that excludes "others." The COVID-19 pandemic, to draw on Pierre Bourdieu's (1977) conceptual vocabulary, has ushered in a crisis that reflects further accentuation of the heterodoxy and the orthodoxy in relation to the dominant discourse of the nation that conflates whiteness with American citizenship and nationhood.[76]

The COVID-19 pandemic has become the site of imagining and contesting nationhood not only in the United States. In India, where the present Bharatiya Janata Party Government has sought and actively undertaken policy measures to "imagine" Indian nationhood through Hindu identity, widespread negative coverage of a Tablighi Jamaat (an Islamic sect) event in Delhi in March 2020 became yet another exemplification of the exclusionary imaginary of India. The media coverage of this event, a study concluded, "was demonstrably biased" and "put the blame on a religious minority," namely Muslims (Sharma and Anand 2020). An extensive study of people's views in 40 neighborhoods in Uttar Pradesh, India's most populous state, found that "93% of participants believed that foreigners were responsible for the ongoing disaster [in relation to the COVID-19], with a further 66% blaming Muslim populations for the spread of COVID."[77] These data are again reflective of not simply contingent viewpoints of people or of biased media coverage, but of an "imagined community" that seeks to include/exclude who belongs to the nation, further reinforced through governmental strategies that have a colonial genealogy (see, e.g., Banjerjee 2020).

In the context of Brazil the pandemic revived "painful memories and well justified fears" of European colonization of the Americas that had led to "several Amerindian groups" being "completely wiped out by exogenous diseases like measles and smallpox" (Charlier and Varison 2020: 1069). President Jair Bolsonaro's response to the pandemic, which has been called an "institutional strategy for the spread of coronavirus," along with his "anti-indigenous policies," it has been argued, "amounts to . . . a genocide against Brazil's first peoples."[78] Ironically, the colonial history of Brazil/Americas combined with Bolsonaro's anti-COVID protection measures, including his claim that "[n]obody can force me to get the vaccine," has magnified the misgivings and misinformation among the indigenous communities about the vaccine. So much so that a group of "villagers, part of the indigenous Jamamadi group, greeted" a "helicopter loaded with health workers and coronavirus vaccine doses" "armed with bows and arrows – and demanded that it leave."[79] The recent pandemic, in short, has become an important site to, explicitly or implicitly, cultivate and contest nationhood that has commonly been imagined through the identity of the dominant social group. In the next and the last section of this chapter, I focus on claims of SARS-CoV-2's origin in China and investigate its entanglement with the concern to maintain global order.

Situating the origin of COVID-19 and maintaining global order

Obama was a traitor.
America, he hate her.
He belongs to jails.
I ain't lying, it ain't no jokes
Corona is a liberal hoax.
Obama, what we are gonna do?
Inject him with Wuhan Flu.
Dr Fauci, what we gonna do?

Inject him with Wuhan Flu. [microphone is turned
  towards the audience and they sing this part]
  . . .
          The character Borat, a journalist from Kazakhstan,
                  in *Borat Subsequent Moviefilm* (2020)

This quote, which is part of a song, is from a comedy
film that is shot as a documentary through a fake character, Borat, who masquerades as a journalist from
Kazakhstan and together with his daughter (another
character in the film) infiltrates and documents the views
of some conspiracy groups and a few political leaders in
the United States. The song reiterates commonly used
tropes of "othering" that are fused together and acquire
new meanings in the context of COVID-19. Wuhan flu,
for example, became a commonly used trope, particularly among Republican and conservative groups, after
SARS-CoV-2 started to infect and kill a large number of
people in the United States. Former President Trump,
during his presidency and even after leaving office,
made referencing the Wuhan flu and the Chinese virus
(even Kung flu) the rallying call for his supporters – he,
for example, "used the expression 'Chinese virus' more
than 20 times just between March 16 and March 30,"
when the COVID-19 infections and deaths surged in the
United States.[80]

Although the then President Trump stated that such
references were "not racist at all," their deployment
ushered another iteration of the "yellow peril," and
Asian Americans experienced racist abuse and violence
throughout the pandemic.[81] A study of tweets that were
posted in two weeks in March 2020 (starting a week
before and ending a week after President Trump's tweets
calling COVID-19 the Chinese virus), which was published in the *American Journal of Public Health*, showed
that while nearly 20% (19.7%) of the "hashtags with
#covid19 showed anti-Asian sentiment," the figure for

such a sentiment jumped to more than "half (50.4%) of the ... hashtags with #chinese virus" (Hswen et al. 2021: 956). Hashtags on Twitter such as #burnwuhan, #bombchina, #chinaliedpeopledied, #nukechina, as the above-quoted study showed, often accompanied anti-Asian comments.

The conspiracies centered on the origin of COVID-19 regained momentum after President Joe Biden, a Democrat, instituted an investigation into the possible Chinese origin of the SARS-CoV-2 virus. President Biden, highlighting that "the origins of COVID-19" could be "from human contact with an infected animal or from a laboratory accident," stated that the "United States will ... keep working with like-minded partners around the world to press China to participate in a full, transparent, evidence-based international investigation."[82] The US media went into overdrive to show how the lab leak theory is plausible after President Biden's statement. The *Wall Street Journal* published two articles, one of which was titled "Intelligence on Sick Staff at Wuhan Lab Fuels Debate on Covid-19 Origin."[83] This article, referring to "an undisclosed document from an anonymous official who was a part of the former US president Donald Trump's administration," suggested "that 3 WIV [Wuhan Institute of Virology] employees were sick in November 2019."[84] Instances like these, which had been denied earlier and there was no additional evidence to suggest to the contrary, became the basis of the COVID-19 lab origins claim.

Vanity Fair went a step further and cited a data scientist based in New Zealand, who had started independent research on the COVID-19 origins after reading the statement from 27 scientists that was published in the *Lancet* (Calisher et al. 2020), which had "strongly condemn[ed] conspiracy theories suggesting that COVID-19 does not have a natural origin." For this data scientist, the statement published in the

Lancet seemed as though "nailed to the church doors": "Everyone had to follow it. Everyone was intimidated. That set the tone." The *Lancet* statement, according to this data scientist, was thus *"totally unscientific"* (emphasis added). Again, no new and compelling evidence for the lab origins was presented; nevertheless, the Chinese responses, according to the *Vanity Fair* article, "smelled like a cover-up."[85] The evidence against the lab origins of SARS-CoV-2, including statements by the Australian virologist Danielle Anderson, who was working at the Wuhan Institute of Virology in November 2019, was sidelined in the process. In fact, after Dr Anderson, in collaboration with another scientist, "declared the claims [published in an article in the *New York Post*] to be misleading," she "had her name trashed so viciously by extremists she had to call in police."[86]

Interestingly, President Biden does not refer to science or scientists even once in his statement that was released by the White House on May 26, 2021. Scientists, who by and large had argued against the possible lab origins of SARS-CoV-2, were, however, also involved in the resurgence of the Chinese lab-origin discourse. A few days before the White House released President Biden's statement calling for an investigation, several scientists published a letter in *Science* that was titled "Investigate the Origins of COVID-19" (Bloom et al. 2021). This particular call for investigation quickly became widespread and provided legitimacy to a claim that was largely associated with conspiracy theorists and had remained a peripheral concern for a large section of scientists (again, we can see how the credibility of science and scientists can fuel misinformation and conspiracies).

These scientists' call for an investigation into the possible lab origins of COVID-19 that was published in the journal *Science* is telling. It states: "As scientists with relevant expertise, we agree with the WHO director-general, the United States and 13 other countries, and the European Union that greater clarity about the origins of this pandemic is necessary and feasible" (Bloom et al. 2021: 694). The claim of these scientists reads like a diplomatic brief. To highlight the support of "the United States and 13 other countries" this article, for example, cites a statement issued by the US State Department in March 2021. The claim of these scientists is also misleading. The WHO director-general, Dr Tedros Adhanom Ghebreyesus, who as Reuters had claimed got "caught in Trump–China feud" and was "under siege" throughout the pandemic,[87] in his remarks was hardly emphatic about such an investigation. Dr Ghebreyesus stated that "[a]lthough the team has *concluded that a laboratory leak is the least likely hypothesis*, this requires further investigation, potentially with additional missions involving specialist experts, which I am ready to deploy" (emphasis added).[88]

Responding to the resurgence of calls for investigation of the lab origins of COVID-19, an article published in *Nature* argued that the evidence presented as indicative of a possible lab origin of the virus could equally well be witnessed in the context of a natural origin.[89] "David Relman, a microbiologist at Stanford University in California," who is the lead author of the statement calling for an investigation of the lab origins of COVID-19 in the journal *Science*, stated, "I am not saying I believe the virus came from a laboratory." According to him, "the authors of the WHO investigation report *were too decisive in their conclusions*" (emphasis added).[90] Relman is not asked what additional evidence led them to call for such an investigation, considering, among other known facts, it is also believed that "lab leaks have never caused an epidemic, [though] they have resulted in small outbreaks."[91]

I am not suggesting that the concern of these scientists is misplaced nor arguing that there is no possibility of a lab origin of the virus. I simply wish to highlight that the concern about the origin of SARS-CoV-2 raised by scientists in the reputable journal *Science* is forked and explicitly entangled with the concern of several dominant western countries with the rise of China. I use the word "forked" here to also highlight the colonial genealogy of such discursive emplotments. Specifically, as Homi Bhabha, drawing on a Native-American proverb, argues, the "discourse of post-Enlightenment . . . colonialism often speaks in a tongue that is forked, not false" (Bhabha 1994: 122).

The politics of the COVID-19 lab origins and the concern with the rise of China were front and center at the June 2021 G7 meeting. US President Joe Biden left little in doubt in this regard when he stated "I think we're in a contest with China . . . in a contest with autocratic governments around the world, as to whether or not democracies can compete with them in a rapidly changing twenty-first century."[92] The anxiety behind this concern, and also its irony, is evident in the striking figures of COVID-19 infections and deaths in the G7 countries. Of the total of 197 million infections and 4.2 million deaths until July 2021, 57.9 million infections and 1.13 million deaths had occurred in the G7 countries.[93] That is, the G7 countries, which together constitute nearly 40% of global gross domestic product (GDP) but have about 10% of the population of the world, have had more than a quarter of the COVID-19 infections and deaths. In several so-called "Third World," "underdeveloped," and "developing" nations, the large number of COVID-19 infections and deaths in countries such as the United States and the United Kingdom became an excuse to justify the failures of their own governments in relation to COVID-19.

In contrast, although the pandemic started in China, it had not only managed to control the spread rather effectively (even former President Trump had praised the early efforts of China) but it had also been providing vaccines and other medical support to countries around the world. Before the latest reinstatement of concern with China's role in COVID-19 origin, western media had been highlighting how, for example "vaccine diplomacy . . . [was boosting] China's standing in Latin America."[94] And then, at the time when the origin of virus controversy was reignited, it was also being headlined how the "US . . . [was starting to blunt] China's vaccine diplomacy in Latin America" by starting supply of vaccines to South American countries.[95] In short, virological concerns about COVID-19 have been inseparable from statecraft aimed at maintaining a West-centered global order. The entanglement of virology with statecraft and international diplomacy is not unique to the United States. It has been happening globally (e.g., in China, India, Brazil, etc.), and these entanglements reflect the existing global hierarchy and challenges to it.

The concern with the possible lab origins of SARS-CoV-2 also highlights a human-centered and colonial approach towards the management of the pandemic. Bruno Latour has rightly criticized "the noise surrounding a 'state of war' against the virus."[96] A number of biologists have been raising this concern for more than two decades. Joshua Lederberg, who was awarded the Nobel Prize in 1958 "for his discoveries concerning genetic recombination and the organization of the genetic material of bacteria,"[97] while noting that "the realization that they [germs] must inexorably be evolving and changing has been slow to sink in the ideology and practice of the public health sector," had categorically warned: "*Teach War No More*" (Lederberg 2000: 288; emphasis added). Lederberg had added that "our most sophisticated leap would be to drop the Manichaean view of microbes – "We good; they evil" (Lederberg 2000: 288), which, in different avataars, has also been a colonial trope that justified the civilizing missions of the Europeans. Lederberg proposed that, instead of war on pathogens, we should adopt an ecological approach that recognizes that "humans, animals, plants and microbes are cohabitants of the planet" (Lederberg 2000). In 2009, the "Forum on Microbial Threats" of the National Academies, which was dedicated to the "life and scientific legacies of Joshua Lederberg," reiterated the need for an ecological approach (Relman, Hamburg, Choffnes, and Mack 2009).

The ecological approach, as Lederberg had suggested, called for redirection of efforts towards "questions that focus on the origin and dynamics of instabilities within this context of cohabitation," whereby "instabilities arise from two main sources loosely definable as ecological and evolutionary" (Lederberg 2000: 288). Unfortunately, countries across the globe have been stuck in the older model of pathogenesis that is guided by the war metaphor, although SARS-CoV-2 itself is constantly reminding us of the dynamism of its role and the shifts in its form through various mutations (Delta, Lambda, and so on). China's zero-COVID policy, after it aggressively controlled the earlier phases of infection, is the most striking example of the war on pathogen approach, and it mimics the colonial strategies, albeit applied to the virus and Chinese citizens.[98] The call for an investigation of the COVID-19 origins in China that was raised by the scientists in *Science* (Bloom et al. 2021) translates this approach for international statecraft.

The statement of the scientists, which was published in *Science*, also highlights why the claim that anti-science beliefs are hampering effective action even in cases of widespread scientific consensus, for example in relation to climate change, may be only partially true. A few years ago, Bruno Latour, one of the founders of the field of science and technology studies (STS) and one of the best-known scholars in the field, was described in an interview in *Science* thus: "Bruno Latour was a thorn in scientists' sides. Now, he wants to rebuild trust in their work." Latour was specifically asked, "How should scientists wage this new war [against 'the rise of antiscientific thinking' and 'alternative facts']?" He replied that "[w]e will have to regain some of the authority of science," adding, "[t]hat is the complete opposite from where we started doing science studies." Nevertheless, for Latour, "the solution is the same: You need to present science in action" (Vrieze 2017: 159).

As I have argued, trust in science and scientists has not decreased. In fact, it remained very high even during the pandemic when we witnessed widespread protests against vaccines and the mask mandate. Moreover, as I also showed, conspiracy theorists and misinformation spreaders often used the credibility of science and scientists to make their claim. Further, in the war against "anti-science" thinking, we need to consider the role of scientists themselves in spreading misinformation and conspiracies. For example, the French virologist Luc Antoine Montagnier, who had shared the Nobel Prize in physiology and medicine for the discovery of the human immunodeficiency virus (HIV) in 2008, had written the blurb for Judy Mikovits's book, *Plague of Corruption* (Mikovits and Heckenlively 2020), thereby sharing a platform with well-known anti-vaxxers such as Kent Heckenlively and Robert Kennedy Jr. Robert Kennedy Jr wrote the preface to Mikovits's book, in which he compared Mikovits to Galileo. Not to forget, as I have also shown, Mikovits, who became an important node in the spread of COVID-19 misinformation through the video *Plandemic*, is herself a trained scientist with a PhD. Montagnier was also among the first to claim that "COVID-19 was made in a lab" – a much stronger claim than that being made by the Biden administration and some other scientists.[99] Do these scientists manifest "anti-science thinking"?

Let me categorically state that I am not suggesting, as has been commonly argued, that opposition to, say, vaccines or climate change research is not about science (but about politics, economic interest, etc.). *Rather, deployment of the term "science" – in the singular – has always been and continues to be entangled with elements of knowledge, identity, politics, economic interests, and so on.* That is, while making claims in relation to, for instance, COVID-19, it is one thing to say that studies in virology (and those from other related disciplines) show how such and such claim is true or false. But what does it even mean when we characterize a claim as science or anti-science? The term "science" acquires specific meaning and reference only when presented in relation to the "other" – superstition, magic, primitive, the non-West, etc., which, as I argue in chapter 2, has been tied to Orientalism and colonialism. We cannot, thus, defend "science" without decolonizing it from such multilayered and hierarchical meanings and references.

2

Historicism without History: The Scientific Revolution, Reimagining the European Past, and Postcolonial Futures

> The Scientific Revolution that emerged in northern Europe during the seventeenth century arrived in Latin America along with programs for economic and cultural renewal launched by Charles III in Spain and the Marquis of Pombal in Portugal.[1]
>
> Jorge Cañizares-Esguerra and Marcos Cueto, "Latin America Science: The Long View," *nacla*, September 25, 2007.

> Africa's scientific revolution must start at the roots.
>
> Editorial, *SciDev.Net*, February 1, 2007.

The Scientific Revolution is arguably the most well-known and widely accepted historical event globally. Be it news media or textbooks for schoolchildren and college students, the concept of the Scientific Revolution has come to have, broadly, a self-evident meaning and reference all over the world. A textbook for BEd (bachelor's degree in education) in India, published by the National Council of Educational Research and Training (NCERT), for example, defines it thus: "The scientific revolution triggered by Copernicus and steered vigorously ahead by Kepler and Galileo was brought to a grand completion by Newton . . . *The age of reason had dawned*" (emphasis added).[2] And as the quotes in the epigraphs highlight, this paradigmatic historical shift has been considered not simply a local event whose influence was restricted to Europe but a model of the future for the rest of the world.

The Scientific Revolution as denotative of a historical change (rather, a rupture, as the word "revolution" signifies) that marked the birth of a new way of thinking has become so commonplace that it is difficult to even imagine that the concept was first used in this sense in the 1940s (Butterfield 1957; Koyre 1943).[3] In particular, Herbert Butterfield's lectures at Cambridge University in 1948 "in published form proved to be extraordinarily influential, deeply shaping the work of historians of science in the postwar years" (Westman and Lindberg 1990: xvii). "The Scientific Revolution," as Margaret Osler, writing around fifty years after Butterfield's lectures, states in the introduction to *Rethinking the Scientific Revolution* (Osler 2000b), "is probably the single most important unifying concept in the history of science" (Osler 2000a: 3). It has spawned a vast literature in not only the history of science but a number of academic fields. This literature also includes reappraisals (see, e.g., essays in Lindberg and Westman 1990) and rethinking (see, e.g., the essays in Osler 2000b). These reappraisals and rethinking, while adding new historiographic interventions, reinforce the Scientific Revolution as a "canonical imperative."

Despite its canonical status, Steven Shapin rightly suggests that "historians have become increasingly uneasy with the very idea of 'the Scientific Revolution'" and "the legitimacy of each word making up that phrase has been individually contested" (Shapin 1996: 3). The very first sentence of his book declares "There was no such thing as the Scientific Revolution" (1996: 1). Nevertheless, Shapin argues for the need to write about the Scientific Revolution, firstly, because "many key figures in the late sixteenth and seventeenth centuries vigorously expressed *their* view that they were proposing some very new and very important changes in knowledge of natural reality and in the practices by which legitimate knowledge was to be secured, assessed, and communicated" (1996: 5). And, secondly, for Shapin, "the very idea of the Scientific Revolution . . . is at least partly an expression of 'our' interest in our ancestors, where 'we' are late twentieth-century scientists and those for whom what they believe counts as truth about the natural world" (1996: 6).

I am sure historians of science as well as scholars from a range of other disciplines will continue to engage with the Scientific Revolution and expand our understanding of that period of European history. In this chapter, building on Marwa Elshakry's claim that "in many ways the history of science itself started off by asking if science was the specific product of Western civilization" (Elshakry 2010: 99), I seek to contribute to the vast and ever growing number of studies of the Scientific Revolution. Specifically, I focus on the Eurocentric historicism that the Scientific Revolution enfolds. This historicism, to draw on Dipesh Chakrabarty's argument, is articulated through the temporal order of "first in Europe, then elsewhere" and relegates the non-West to the "waiting room of history" (Chakrabarty 2000). Eurocentric historicism, which Chakrabarty sought to provincialize, posits "historical time as a measure of the cultural distance . . . that was [and is] assumed to exist between the West and the non-West" (Chakrabarty 2000: 7). Chakrabarty was not suggesting an abandonment of European ideas and thought. His "solution is to renew European thought from the margins – to 'provincialize Europe'" (Satia 2020: 266).

Chakrabarty's concern with provincializing Europe has acquired an almost cult-like status, wherein just its citation is seen/shown as evidence of an anti-Euro-/West-centric position (see chapter 3). However, his book also gave birth to vigorous engagements with Eurocentric historicism. Priya Satia, emphasizing Chakrabarty's argument that "historicism enabled European domination of the world in the nineteenth century," for example, engages with a vast corpus of historical studies and European philosophizing of the role of history and goes on to show "*how* historicism did this" (Satia 2020: 3). My focus in this chapter is simpler and much narrower. I intend to show how the concept of the Scientific Revolution enfolds a historicism without history that made, and continues to make, "modern science" an ideal model to imagine and enact futures for the non-western as well as western world. On the one hand, science became "pivotal in the imagination and institution" of newly independent nation-states (Prakash 1999: 3) and, on the other, this Eurocentric historicism also became a model for seemingly anti-science historical imaginaries (Prasad 2018).

In the slightly more than two decades that have followed the publication of Chakrabarty's book, the global geopolitical power has dramatically shifted and, among other things, it has spawned historiographic interventions that mimic Eurocentric historicism but overturn the European/western hierarchy. "Proponents of Hindu science," for example, in contrast to the usual argument that they are anti-science, "present achievements of ancient India's past . . . as a direct challenge to the accepted histories of inventions and discoveries," and in doing so "caricature the Eurocentric structure to present an Indo-centric historicism – first in India then elsewhere" (Prasad 2018: 55–6; Subramaniam 2000). In short, what we are witnessing is not necessarily science versus anti-science, but competing histories of *science*, although the Euro-/West-centric one not only remains dominant but is also the historiographic model for others. The issue then for me is not decentering Europe/

the West, but rather contributing to a discursive clearing that can allow us to recover connected and entangled histories of sciences that do not entrap us within various civilizational and societal silos through dualistic categories that we owe to European modernity (Prasad 2014).

Another reason for my critique of the Eurocentric historicism inherent in the conceptualization of the Scientific Revolution is to highlight its central role in constituting science as a singular and universal culture that not only transcends particular social influences but is projected as a particular state of being – as being scientific. Indeed, as I show in this chapter, that is what allows Eurocentric historicism to be effective. However, this idealized imaginary of science can also become a tool to spread misinformation and conspiracies, as I showed in chapter 1 in the context of the COVID-19 pandemic. The very use of the term "Scientific Revolution" in the singular signifies such a reference. The historiography of the Scientific Revolution sharpened and reinforced the idealized imaginary of a singular and universal science and projected it also as a way of life. As Yehuda Elkana has argued, Koyre's "studies . . . in the history of scientific thought in the sixteenth and seventeenth centuries, the 'Scientific Revolution' . . . became a paradigm for *history of science as history of disembodied ideas*" (Elkana 1987: 111; emphasis added). The understanding of science that undergirded Koyre's and Butterfield's historical arguments about the Scientific Revolution was to a large extent already charted by the historians of science preceding them. George Sarton, among whose enormous contributions is the establishment and editorship of *Isis*, which is still the best-regarded journal in the history of science, played a pivotal role in this regard.

The rest of the chapter is divided into three sections. In the first section, I show how the history of science, particularly as it was crafted by George Sarton's efforts, posits a singular and universal science that is argued to be a model not only of scientific practice but also for being a "scientific" person. The second section focuses on Herbert Butterfield's historicizing of the Scientific Revolution, which, as I discussed earlier, has been deeply influential. I argue that his historiographic articulation of the Scientific Revolution was dependent on the "othering" of the non-West, particularly Asia, that not only undercut any role for the non-West in the emergence of "modern science" but also presented the non-West as a hostile and aggressive "other." This "othering" of the "Asiatics"/non-Europeans undergirds the Eurocentric framing of the Scientific Revolution. Thereafter, in the last section, I analyze Joseph Needham's civilizational approach to understanding the birth of modern science, which represents a version of the divergence theory (Raj 2016). Needham's historiographic arguments, in contrast to Butterfield's, highlighted the constitutive role of non-western techno-scientific developments in the eventual emergence of modern science in Europe.[4] Although Needham's historical accounts were largely positive towards non-western others, in particular China, his historiography of the non-western "other," as I show, complements Butterfield's Eurocentric characterization of the Scientific Revolution. In simple terms, Needham's ecumenical approach made the Eurocentric historicism, inherent in the concept of the Scientific Revolution, if not palatable, at least something that could be ignored or was not in need of direct challenge.

Science, history, and Europe/the West

Aristotle is considered by many to be the first scientist, although the term postdates him by more than two millennia.

Roberto Lo Presti, "History of Science:
The First Scientist," *Nature*, 2014: 250

Although Sarton's Comtean view of the history of science as a saga of intellectual and moral progress that should evaluate the past in light of the present scientific understanding ... has left little trace in current historiography ... his tireless attempts to institutionalize and organize the discipline ... had a lasting impact on the field.

> Lorraine Daston, *International Encyclopedia of the Social & Behavioral Sciences* (Second Edition), 2015: 242

A "century ago," Peter Dear points out, "scientists tended to represent themselves as participants in a historically ongoing cultural enterprise" (Dear 2009: 90). Joseph "Larmor [in 1910–11] thus provides an up-to-date overview of ideas, theories, and problems in relation to the aether that takes the form of a historical narration; *his history is constitutive of his scientific account*" (2009: 90–1; emphasis added).[5] In contrast, Sarton's goal, which became the focus of *Isis*, was to establish the *history of science* and not the "history of sciences," which was, for him, quite different" (2009: 91), because the latter represented historically situated "science in the making" that, for example, Larmor published as his research, around the same time as Sarton proposed his conception of the history of science. Sarton's concern was that "science which is making itself ... writes its own history – a provisional history," while "when we speak of *history of science*, what must be understood is the history of science becoming classic ... which constitutes, or ought to constitute, the intellectual baggage of every cultivated man" (as quoted in Dear 2009: 91).

Dear argues that "[s]cience 'making itself' ('*la science qui se fait*'), distinguished from 'history of science' ('*histoire de la science*'), resembles Bruno Latour's well-known distinction between 'science in the making' and 'ready-made science'" (Dear 2009: 91). There is, however, a radical difference between Latour's distinction and that made by Sarton. Latour's distinction is limited to the practice of sciences wherein the settling of disputes through a network of allies results in transforming "science in the making," or techno-science, to "scientific facts" (Latour 1987, 1999, 2005). The concern for Sarton, which can be seen as an exposition of the "modern scientific culture," was the transformation of being/self as well as the society and civilization through science – as *the intellectual baggage of every cultivated man.*" That is to say, when we hear and use phrases such as people being "anti-science" or "the need to be scientific" (see chapter 1), we will have to recognize that such an understanding of science – as a state of being – is an artifact of the history of science, albeit Sarton and other historians of science may not be the first to use the terms "science" or "scientific" in that sense.[6]

In an essay published in *The Monist* in 2016, Sarton further explained "the meaning of the history of science, to determine its limits and to show how it should be studied" (Sarton 1916: 321). Sarton stated at the outset: "The history of science is the study of the development of science – just as one studies the development of a plant or an animal – from its very birth" (1916: 321). As one can see, the very idea of history of science as a discipline, which itself is being born through the engagements of Sarton and others at this time in history, is tied not just to understanding *science* (in the singular), but to the investigation of the birth and development of science.[7] Sarton elaborates a little later in the article, "[t]o *secure the unity of knowledge* it will be more and more necessary that some men make a deep study of the principles and of the historical and logical development of all the sciences" (Sarton 1916: 330; emphasis added).

Sarton was also clear that the history of science is "a history of human civilization, considered from its highest point of view" (1916: 333). The history of science, thus, not only gave "birth" to an idealized science, but also presented science as a source and an embodiment of a "new humanism" (see also Sarton 1924). In the words of Sarton: "*The history of science, if it is understood in a really philosophic way, will broaden our horizon and sympathy*; it will *raise our intellectual and moral standards*; it will *deepen our comprehension of men and nature*" (Sarton 1916: 357; emphasis added). Thus, "[i]t is the historian's duty to evidence all the scientific facts and ideas that make for peace and civilization; in this way he will better secure science's cultural function" (1916: 359). "The history of science," according to Sarton, "is accomplishing an endless purification of scientific facts and ideas" (1916: 350).

Science defined as such, which became the object of the history (and also the philosophy and the sociology) of science, was not to be seen as a local practice. Sarton cast the practice of science as global cosmopolitanism, in a sense prefiguring studies such as that of Michael Polanyi on the "republic of science," although Polanyi was a critic of positivism, which Sarton espoused (Polanyi 1951, 1962; Polanyi, Ziman, and Fuller 2000). "Scientific work," Sarton argued, "is the result of an international collaboration, the organization of which is perfected every day," adding: "Science aims at objectivity; the scientist exerts himself to decrease to a minimum his 'personal equation'" (Sarton 1916: 341).

The reason for my exegesis of Sarton's arguments about the role and function of the history of science is not to suggest that he is the sole "author" of the discipline or that his arguments have been accepted without any challenges or modifications. The wide-ranging influence of Sarton and his work cannot be denied. Writing on the centennial of Sarton's birth, Eugene Garfield not only called Sarton "the father of the history of science," who apart from his monumental works in history of science had also held the editorship of *Isis* for forty years, but also emphasized how Sarton had "come to epitomize the history of science to scholars throughout the world" (Garfield 1985: 125). Sarton and also Butterfield and Needham, for me, to draw on Michel Foucault's argument, represent "'author' as a function of discourse" (Foucault 1977b: 124) That is, "they produced not only their own work, but the possibility and the rules of formation of other texts" (1997b: 131). Specifically, the historiographic elaboration of a particular understanding of science and of the Scientific Revolution by these authors constitutes the foundational grid of Eurocentric historicism, which is the focus of this chapter. Nonetheless, it needs emphasizing that these authors are themselves part of a long genealogy of western imagination of superiority of the self through science.[8]

Indeed, as Michael Adas has shown, there is a long history of European administrators and scholars deploying "the centrality of scientific and technological accomplishments as proof of the superiority of Europeans over non-Western peoples" (Adas 1989: 194). By the early nineteenth century, scholars/administrators such as James Mill also started "to link [the European] material achievement to a new sense of what it meant to be civilized" (Adas 1989: 194). Nevertheless, the meaning and reference of "modern science" was not fixed and they continued to be syncretically produced, both in the West and the non-West, although much of the globe had been colonized by the Europeans (Elshakry 2008, 2010; Fan 2007a; Raj 2007). In different non-western societies/nations, depending upon the particulars of the colonial encounters with the West – for example, the Opium Wars in China or World War I in Japan – the meaning and reference of "modern science" got reconstituted (see, e.g., Elshakry 2010; Mizuno 2009).

It is important to note that "the creation of a notion of a universalizing 'western science' helped lay the conceptual foundation for 'Chinese science,' 'Arabic science,' and even 'Islamic science,' among other similar categories" (Elshakry 2010: 106–7). To draw on Elshakry again, "just as the disciplinary history of science gave birth to the ideal of a singular and unified 'science' in academies in Europe and America, others around the world were beginning to institutionalize their own local versions of a more plural story" (Elshakry 2010: 107). The only thing I would like to add to Elshakry's claim is that the historiography of the Scientific Revolution that emerged around World War II and gained currency globally thereafter made the local versions of "modern science" manifestations of lag and/or lack in relation to the "imaginary" of the universal modern science, which was, nevertheless, seen/shown as western.

The idealized imaginary of science produced through the history of science (and through the works of philosophers, sociologists, and scientists as well – Needham, for example, was trained as a chemist and not a historian) established, as Sharon Traweek famously described in the context of particle physics research, science as a "culture of no culture" (see also Franklin 1995; Traweek 1988). However, the Scientific Revolution that is argued to mark the birth of such a science, as I briefly discussed in the Introduction, enfolds a forked articulation. Modern science came to be articulated simultaneously as a culture of no culture and a product of the European/western culture, and this was achieved through the deployment of the term "revolution" (a term marking a radical rupture, thereby presenting the birth almost like an immaculate conception). What we have as a result is *Eurocentric historicism without history* that was presented, and also enthusiastically adopted, as an ideal model for the soon to emerge postcolonial nations.

This forked articulation of modern science – as European in origin and yet embodying a universal culture – became the basis for the diffusion model of science. Indeed, as Bruno Latour argues, *"belief* in the existence of a society separated from technoscience *is an outcome of the diffusion model"* (Latour 1987: 141). Nonetheless, the diffusion model has also been umbilically tied to Europe/the West. We need to additionally consider whether the *diffusion model is itself an outcome of Eurocentric historicism.* George Basalla, in his widely cited and also criticized article, for example, wrote at the outset: "The relatively small geographical area covered by these nations [of Western Europe] was the scene of the Scientific Revolution *which firmly established the philosophical viewpoint, experimental activity, and social institutions we now identify as modern science"* (Basalla 1967: 611; emphasis added).[9] Basalla did not mention what were the specific elements of the philosophical viewpoint or experimental activity that were born during the Scientific Revolution; simply a reference to modern science and its European birth made the meaning and constitution of these categories seem self-evident. Basalla could, thus, shift to discussing how science diffused to the non-West. This idealized imaginary of modern science, and with it of scientific culture, was not forced on the non-western people and nations (although its "currency" in the non-West cannot be extricated from European colonialism). It was, and continues to be, sought and widely accepted in non-western societies.

The concern in relation to models such as the one proposed by Basalla is not simply about the particularities of diffusion of science to the non-West. As I have shown elsewhere, "the *premise of diffusion of universal, western science to non-western cultures* is what gives credence to the claim that Western Europe is the exclusive point of origin for modern science" (Prasad 2019: 1078; emphasis added). As such, the specific elements of modern scientific culture and practices come to have meaning as a lack in the "other" that can be overcome if the lag in scientific knowledge, culture, and so on from the West is bridged. What follows then is a never-ending discourse of lag and lack that are sought in the non-West and continually expressed through various characteristics (lack of scientific institutions, lack of scientific culture, and so on), which, like Homi Bhabha's (1994) description of the colonial stereotype, become "metonyms of presence."

Europe/the West, as a result of this colonial stereotyping, becomes an embodiment of the "scientific culture" that the non-western "other" is argued to be lacking. Such stereotyping is, for example, strikingly evident in the former United States' Secretary of State Henry Kissinger's understanding of the global order and international diplomacy: "The West is deeply committed to the notion that the real world is external to the observer, that knowledge consists of recording and classifying data – the more accurately the better. *Cultures which escaped the early impact of Newtonian thinking have retained the essentially pre-Newtonian view that the real world is almost completely internal to the observer*" (Kissinger 1977: 48; emphasis added; see also Said 1979).

Kissinger, who received the Nobel Prize for Peace in 1973, goes on to argue: "Technology comes as a gift; acquiring it in its advanced form does not presuppose the philosophical commitment that discovering it imposed on the West." "Empirical reality has a much different significance for many of the new countries than for the West" as a result (Kissinger 1977: 49). This contemporary articulation of "machines as the measure of men" (Adas 1989), bestowed as a "philosophical schism," according to Kissinger, was at the core of the "instability of the contemporary world order" (Kissinger 1977: 49).

One may argue that the views of, for example,

Kissinger on global order and the concerns of the history of science are not directly connected. However, that would be to forget Edward Said's (1979) forceful analysis of Orientalism's complex entanglements across different domains of knowledge and practice in the West. In fact, as I show in the next section, Butterfield also invokes Newton while highlighting the shift that came about during the Scientific Revolution in Europe/ the West, although Newton was born a hundred years after the death of Copernicus (whose heliocentric view is seen as marking the birth of the Scientific Revolution). Such historiographic slippage and compression of historical time has been a part of Euro-/West-centric strategy to define Europe's history as well as identity through modern science.

In the next two sections, I focus on the historiography of the Scientific Revolution around World War II, which was also the beginning of the end of formal European colonialism. In the next section, I analyze Butterfield's history of the "origins of modern science" in Europe and thereafter focus on Joseph Needham's civilizational approach to explain the birth of modern science. I argue, as stated earlier, that Needham's and Butterfield's approaches, whose influence continue in the recent histories of sciences as well as in the western imaginary, were complementary and gave birth to a Eurocentric historicism without history.

Butterfield, the Scientific Revolution, and reimagining of the European past

[The Scientific Revolution] outshines everything since the rise of Christianity and reduces the Renaissance and Reformation to the rank of medieval Christendom.

Herbert Butterfield, *The Origins of Modern Science*, 1957: 7

> We need to know how man came to acquire a concept
> of "the past," and gradually to clarify that concept and
> endow it with a structure.
>
> Herbert Butterfield, *The Origins of*
> *History*, 1981: 14

Herbert Butterfield was already a well-known historian, particularly because of his influential book, *The Whig Interpretation of History* (Butterfield 1965 [1931]), when he gave the famous lectures at the University of Cambridge in 1948, which, as Robert Westman and David Lindberg argued, "offered a 'general historian's' view of a cultural change that he referred to as the Scientific Revolution" (Westman and Lindberg 1990: xvii). Although the phrase "scientific revolution" was used prior to Butterfield's lectures, and Alexandre Koyre had deployed the concept to signify the paradigmatic shift in the way to view nature with Galileo's "geometry of motion, [that represented] *a priori*, mathematical science of nature" (Koyre 1943: 347), Butterfield's lectures have arguably been central to ushering in not only a field of historical inquiry, but also a historical and sociological imagination of science and of Europe. The power of the "big picture" of the history of science, Andrew Cunningham and Perry Williams wrote in 1993, "is revealed by the fact that *The Origins of Modern Science* is still in print ... and that students continue to have their first and most formative encounter with the subject either through this book or through others that rely upon variously modified versions of the same big picture" (Cunningham and Williams 1993: 407–8).

In their introduction to the *Reappraisals of the Scientific Revolution* (1990), Westman and Lindberg do not disturb this "big picture" of the history of science. After highlighting Butterfield's influential role, in the very next paragraph they shift to the seventeenth century, pointing out: "Seventeenth-century writers themselves deployed a rhetoric of 'newness' in the titles of their works – Bacon's *New Atlantis* and *New Organon*, Kepler's *New Astronomy*, and Galileo's *Two New Sciences*," for example – and clearly set themselves up in opposition to the orthodoxy of "the schools" (Westman and Lindberg 1990: xvii). They emphasize that "the notion of change employed by Hobbes and his contemporaries did not have the far-reaching meaning of the term 'Scientific Revolution' popularized by Butterfield and Westfall," and added, "In a voice accessible to specialist and nonspecialist alike, Butterfield's *Origins of Modern Science* conveyed the conclusions of the twentieth-century epistemologists of science" (1990: xvii).

This historical continuity, Lindberg argues in a separate essay in the same volume entitled "Conceptions of the Scientific Revolution from Bacon to Butterfield: A Preliminary Sketch," has an even longer arc. According to Lindberg, "Butterfield's view was an outgrowth and continuation of historiographic traditions and European self-perceptions rooted in fifteenth- and sixteenth-century Italian humanism" (Lindberg 1990: 1–2). He thereafter adds: "When, in the course of the seventeenth century, the new science (or, more exactly, the new natural philosophy) came in for appraisal, that appraisal was powerfully shaped by the historical categories and terminology devised by Renaissance humanists" (1990: 2).

In Westman and Lindberg's historical accounts, from which I quoted above, terms such as "science," "revolution," "humanism," "historiographic tradition," "natural philosophy," and so on slide smoothly across literally *several centuries* of social and epistemological changes. The concept of the Scientific Revolution has become so deeply embedded in our understanding that the historicity of the terms and that of the social practices that those terms represented seem to become almost irrelevant. This is ironic considering historians often remind themselves and others of the need for fidelity to the archive, and this concern is commonly articulated also as the basis of their criticism of "theoretical" arguments. Fidelity to the archival findings is an important and valid concern. Nevertheless, when it comes to the big picture of history of science that is presented through the Scientific Revolution and the birth of modern science in Europe/the West, the historical claims are stretched so far and wide that fidelity to archive seems almost senseless.

Moreover, not unlike Thomas Kuhn's (1970) description of the functioning of a scientific paradigm, "anomalies" are continually adjusted to keep this big picture of history of science going. Hence, when the obvious discontinuity to the conceptualization of the Scientific Revolution arose as a result of the knowledge that amateurs, and not scientists, had produced many of the inventions during the Industrial Revolution in England, it was sought to be explained through, for example, "the maze of ingenuity" (Pacey 1992). The question that we need to ask is why is it that, even when every term of the phrase "scientific revolution" has been contested (Shapin 1996), the Scientific Revolution continues to have such a strong influence not just in the history of science but also in the western imaginary of the "self" and the "other"? An answer could be that modern science is not only deeply embedded in the very idea of the uniqueness of the West but its proposed origin through the Scientific Revolution also allows Western Europe to historically constitute a broader and deeper identity of itself that directly links northwestern Europe to ancient Greece and bypasses its "dark medieval ages," in relation to which, most often, the claims of newness were being presented in the sixteenth and seventeenth centuries.[10]

The Scientific Revolution, according to Butterfield, "changed the character of men's habitual mental operations" and "it looms so large as real origin both of the modern world and of the modern mentality that our customary periodization of European history has become an anachronism and an encumbrance" (Butterfield 1957: 7–8). The scientific transformations ushered in by Copernicus, Galileo, and others going down to Newton, according to Butterfield, culminated in the "Philosophe movement in the reign of Louis XIV" – when although "even Newton could not say what were the causes of gravitation . . . men acquired new habits of mind, new methods of enquiry – almost incidentally . . . they founded modern science" (Butterfield 1957: 171). "The results of the scientific revolution," Butterfield further explained, "were precipitately and hastily translated into a new world-view" (Butterfield 1957: 177).[11]

This transformation "in habitual mental operations," according to Butterfield, was a unique achievement of the West. Indeed, the search for the uniqueness of the West was the driving force for Butterfield's historical study of not only the origins of modern science but also of the origins of history itself. Butterfield, in *The Origins of History* (Butterfield 1981), which is presented as the "distillation of the thought and research to which . . . [he] devoted the last twenty years of his life," wrote: "We are often reminded that the civilization of the West is scientific in character; and we do not always remember that it is equally remarkable for being so historically-minded . . . In both fields the development that took place in Europe was unique" (Butterfield 1981: 13).[12]

The uniqueness of the West is, however, defined through and in relation to the non-western "other." Emphasizing the development of scientific character and historical mindedness in Europe, Butterfield argued that "the only known parallel is to be found in the China of comparatively early days, which, besides its amazing feats of science and technology, produced an historical literature of almost incredible vastness." Nevertheless, "in China," according to Butterfield, "there did not develop those modern techniques which, in our section of the globe, led to the scientific revolution of the seventeenth century and the somewhat parallel historiographical revolution of the nineteenth" (Butterfield 1981: 13).[13] In a sense, Butterfield here is presenting his own version of Needham's "Grand Question," albeit extended to historical mindedness as well.

The discursive strategy to explain the uniqueness of the West is evident in Butterfield's historicization of the origin of modern science in Europe, which he elaborated in the chapter titled "The Place of the Scientific Revolution in the History of Western Civilization." Butterfield sought to explain in this chapter "why the West should have come to hold leadership," even though "[u]ntil a period not long before the Renaissance, the intellectual leadership . . . had remained with the lands in the eastern half of the Mediterranean or in empires that stretched further still into what we call the Middle East" (Butterfield 1957: 187–8). During this period, "Constantinople and Baghdad were fabulously wealthy cities, and contemptuous of the backwardness of the Christian West" (1957: 188).

Butterfield's discussion of what he calls the historical shift "in this part of the world" stretches to the Middle East (that became a part of Asia). However, he also calls it "the whole history of Western civilization." He goes on to state, "considering the Graeco-Roman character of European culture in general, it is necessary to account for the division of the continent and to show why there should ever have arisen anything which we could call the civilization of the West" (1957: 188). In Butterfield's historical imaginary of the West, the Middle East continues to be present, albeit as an extension of the Mediterranean. The reference to the East and the West is initially internal but is soon extended to a dichotomous division between Europe/the West and Asia/the East. According to Butterfield, "Even when the Roman Empire surrounded the whole of the Mediterranean, there had been a tension between East and West – a tension greatly increased when a second capital of the Empire was founded and the oriental influences were able to gather themselves together" (1957: 188). The discussion of the West and the East, with the latter referring to the Byzantine Empire, soon acquires a broader geographical scope. Butterfield declares: "From the fourth to the twentieth century one of the most remarkable aspects of the story – the most impressive conflict that spans fifteen hundred years – is *the conflict between Europe and Asia*, a conflict in which down to the time of Newton's *Principia it was the Asiatics who were on the aggressive*" (Butterfield 1957: 189; emphasis added).

Butterfield goes on to argue: "These Asiatic invaders had something to do with the downfall of Rome and the western empire over fifteen hundred years ago" (1957: 189). And, "[b]ecause of their ["Asiatic"] activity over so many centuries it was the western half of Europe that emerged into modern history as the effective heir and legatee of the Graeco-Roman civilization" (1957: 189–90). Butterfield not only attributes the cause of the so-called "Dark Ages" of Europe to the invasions and aggressiveness of the "Asiatics," but also presents an intellectual and historical link between northwestern Europe and ancient Greece. He elaborates: "the course of the seventeenth century ... represents one of the great episodes in human experience, which ought to be placed – along with the exile of the ancient Jews or the building of the universal empires of Alexander the Great and of ancient Rome – amongst the epic adventures that have helped make the human race what it is" (1957: 191). And this is so because the Scientific Revolution occurred in Western Europe at this time. He explains that "[t]here does not seem to be any sign that the ancient world, before its heritage was dispersed, was moving towards anything like the scientific revolution" (1957: 191). "The Scientific Revolution," Butterfield declares, *we must regard, therefore, as a creative product of the West* – depending on a complicated set of conditions which only existed in Western Europe" (1957: 191; emphasis added).

Butterfield's historiography of the Scientific Revolution represents an "imaginative history" (Said 1979) in which historical details become subservient to Eurocentric historicism. In this historiography, not just the Scientific Revolution but through it the identity and history of the West (stretching down to ancient Greece) are framed. And this is achieved by positing a non-western "other" that is shown as the aggressor and a deterrent until the medieval ages, which also allows Butterfield to pass over the so-called "medieval Dark Ages" of Europe. Butterfield's history of the Scientific Revolution, thus, exemplifies Johannes Fabian's argument that the West seems "to require alterity for sustenance in our efforts to assert or understand ourselves" (Fabian 2006: 148).

In this historiography of the Scientific Revolution, the role of European colonialism is all but erased. Its presence becomes at best spectral, seen in references to the European discoveries of various lands across the world (see Butterfield 1957). Interestingly, H. Floris Cohen, who in his book *The Scientific Revolution: A Historiographical Inquiry* sought to present "a comprehensive, critical analysis of the body of . . . literature" on the Scientific Revolution, described the chapter in which, as I have shown, Butterfield presents Eurocentric historicism by "othering" the "Asiatics" "as . . . *one of the very few specifically historically informed treatments of the topic*" (H. F. Cohen 1994: 5; emphasis added). Cohen quotes Butterfield's reference to the European "discovery" of other lands: "The discovery of

the new world, and the beginning of a close acquaintance with tropical countries, release a flood of new data
and a mass of descriptive literature which itself was to
have stimulating effects. *The essential structure of the
sciences was not changed – the scientific revolution was
still far off"* (Butterfield, as quoted in H. F. Cohen 1994:
113; emphasis in the original).

What catches Cohen's attention, which he highlights
in the quote, is the discrepancy in Butterfield's use of the
phrase "Scientific Revolution" because in this instance
the term "revolution" implies "a much narrower time
span to be covered" (H. F. Cohen 1994: 113). The colonizing of the "new world," which resulted in the "great
dying" of the Native Americans (see chapter 3) that
is guised as discovery, is just passed over. The erasure
of colonialism in the historical studies of the Scientific
Revolution, I would argue, makes these studies complicit in the continuing impact of European colonialism
and that of Eurocentric historicism. European empire/
colonialism was certainly not acquired, as argued by the
historian J. R. Seeley, in "a fit of absence of mind" (as
quoted in Satia 2020: 5), but it continues not to have
discursive presence (or to have a distorted or partial
presence) often because of continuing "fits of absence
of mind" on the part of the historians of the Scientific
Revolution.

In Butterfield's historiography of the Scientific
Revolution, the agency of the non-western "others,"
particularly the "Asiatics," is presented as either aggressive or completely passive. Butterfield, for example,
argues that "it would be wrong to impute all the
changes in thought at this time to the effect of the scientific discoveries alone" (Butterfield 1957: 195). The shift
in the "thinking cap" of the Europeans, he suggests, also
happened because "books of travel were beginning to
have a remarkable effect on the general outlook of men
– a postponed result of the geographical discoveries and

of the growing acquaintance with distant lands" (1957: 195; emphasis added). The articulation of these encounters with the non-West is, however, forked and used to present Western Europe's exclusive relation with ancient Greece and to Christianity. "Western Europe was now," wrote Butterfield, "coming to be familiar with the widespread existence of peoples *who had never heard of ancient Greece or of Christianity*" (1957: 195; emphasis added).

In relation to the role of travel books, too, the agency of the non-western "others" is eliminated. Butterfield emphasized that "we have to note that if books of travel affected the very attitude of Western Europeans to their own traditions, the very attitude these people adopted ... owed something to a certain scientific outlook which was now clearly becoming a more general habit of mind" (Butterfield 1957: 196). The stage is, thus, set for the Scientific Revolution to provide the basis for the diffusion of science to the rest of the world – a science that is no longer tied to any culture and presents the Europeans and the West as the only agents of historical change in this regard. Butterfield wrote, "The result was the emergence of a kind of Western civilization which when transmitted to Japan operates on tradition there as it operates on tradition here – dissolving it and having eyes for nothing save a future of brave new worlds" (Butterfield 1957: 202). He added: "It was a civilization that could cut itself away from the Graeco-Roman heritage in general, away from Christianity itself – only too confident in its power to exist independent of anything of the kind" (1957: 202). Butterfield, in short, framed the history of the Scientific Revolution as an embodiment of Eurocentric historicism without history. Needham, as I show in the next section, engaged with the non-western "other," the "Asiatics," very differently. He historically situated various techno-scientific developments in non-western and western societies, which according to him, eventually led to the Scientific Revolution. Nevertheless, Needham remained wedded to the Eurocentric framework of the Scientific Revolution.

Joseph Needham: "The Man Who Loved China"

But Joseph Needham would alter this perception of China [characterized by "a mixture of disdain, contempt, and utter exasperation"] almost overnight and almost single-handedly. Through his many adventures across the country this quite remarkable man would manage to shine the brightest of lights on a vast panorama of Chinese enigmas.

Simon Winchester, *The Man Who Loved China*, 2008: 8

Simon Winchester's biography of Joseph Needham, *The Man Who Loved China: The Fantastic Story of the Eccentric Scientist Who Unlocked the Mysteries of the Middle Kingdom* (Winchester 2008), which was published in 2008 when China had dramatically changed from the time when Needham had written about the Chinese sciences, became a *New York Times* best seller. The book was widely hailed as "captivating," "a fascinating read," and "a masterful biography" in the numerous reviews that were published in reputable newspapers and literary magazines in the United States and other western countries. Winchester, weaving together Needham's personal life with his "adventures" and "discoveries" in China, presents a fascinating account not only of Needham's life and work, but also of China. The narrative, however, unfolds through and within a colonial and Orientalist frame – a white/European man who discovers and unlocks "the mysteries" of the non-western "other." "What matters here," as Edward Said argued in *Orientalism*, "is that Asia [specifically

China in this case] speaks through and by virtue of the European imagination" (Said 1979: 56). Said's argument in relation to Dante's *Inferno* seems quite apt for this biography of Needham as well: "What it is trying to do ... is at one and the same time to characterize the Orient as alien and incorporate it schematically on a theatrical stage whose audience, manager, and actors are *for* Europe [the West]" (1979: 71).

Winchester's book is no doubt a hagiographic biography of Needham and, to use another Said (1979) concept, an example of Orientalist "imaginative history." At one point, for example, Winchester discusses a meeting of Needham with Chairman Mao. In the conversation that is presented as quotes in the book, Mao asks Needham: "You are the only westerner I know well enough, and who knows about motor cars. So I have come to you, Yuese [Needham's Chinese name], to ask for a small piece of advice on an important matter of state business." Mao then asks Needham "whether to allow my people to drive motor cars, or whether the bicycle is better for them. What, my dear Li Yuese, do you think?" Needham responded, according to Winchester, "stuttering slightly": "Well, Mr Chairman ... I find that back in Cambridge where I live, my very old bicycle is perfectly satisfactory for all my needs." Winchester writes thereafter, before Needham could say anything further, "Mao was grinning. He had heard what he wanted to hear," and that enabled Mao to come to a decision. Mao stated: "So, Yuese, you who like China so much find the bicycle perfectly satisfactory ... Right, then. Bicycle it is!" (Winchester 2008: 236–7). Winchester acknowledges after quoting this conversation: "No record of the conversation in Zhongnanhai [between Needham and Mao] exists. Maybe, it never took place" (Winchester 2008: 237).

The fact that Winchester admits that there is no record of such a conversation, most likely, would make academic historians dismiss his claims and perhaps the book itself. Nonetheless, to dismiss Winchester's story for lack of empirical veracity would be to forget that "cultural discourse" circulates not through "'truths' but representations" (Said 1979: 21). Winchester's fiction-alized representation of a dialogue between a Chinese man, albeit the head of the state and one of the most important political figures of the twentieth century, and a British/European scientist is not even mentioned in, for example, the review of the book in the *New York Times*, although the review highlights certain erasures. The review states that "if Winchester's account . . . seems faithful to Needham's character, some careless aspects of the narrative are less so." The reviewer, along with highlighting a couple of repetitions, gives the example of "the map of China accompanying the World War II section," which includes "as-yet-unborn nations like Pakistan, Bangladesh, and North and South Korea."[14] The imaginative excesses of history presented in the biography of Needham, not unlike the narratives of the older Orientalist writings, do not seem to matter to the reviewer.

I start this section with a discussion of Winchester's biography because it allows me to situate Needham within the discourse of Eurocentrism and colonialism and to emphasize two broader issues that I engage with in this chapter and indeed in the whole book. First, as I have also shown in chapter 1, if we were to analyze information/knowledge based only on the veridicality of the claims, we will fail to understand how representations travel. We can, thus, get stuck within the "objectivist fiction" created through dualist divides of science versus non-science or lay perspective versus academic rigor and miss noticing how, for example, as I showed in chapter 1, the idealized imaginary of science becomes a tool to spread misinformation. Second, we would fail to uncover that the acceptance of empirically unproven or wrong information is also a window to the history of the present.

I am not suggesting that veridicality of claims is (or should be) immaterial to historical and sociological studies. My concern is how to make sense of circulation of misinformation (and also information), including colonial imaginative histories such as that of the Scientific Revolution that lack evidentiary proof. We need to pay attention to how, for example, Orientalist and colonial discourses often operate through "objectivist fiction" that is profoundly dependent "on the magic of style to make this trick of truth work" (Taussig 1987: 37).

An important characteristic of European colonialism (and European modernity) has been how Europe/the West has managed to control the production and dissemination of knowledge and information about the "self" and the "other" not just in Europe/the West but across the globe. The western control of knowledge and information flows continues to be pivotal in the maintenance of the hegemony of the West.[15] The concern in this regard is not simply the erasure or control of information. Rather, we need to be alert to how the circulation of information and knowledge about the self and the other became constitutive elements of our identity (in the West as well as in the non-West) and got folded into biopolitical and geopolitical strategies. For example, as Shruti Kapila argues, "the late eighteenth century was . . . marked by the systematization of race theory precisely in the context of the circulation of ideas, networks of imperial institutions, and the inequity of power relations" (Kapila 2010: 123).

Let me return to Needham. Needham is not frequently discussed in historical engagements with the Scientific Revolution. David Lindberg and Robert Westman's edited volume, *Reappraisals of the Scientific Revolution* (Lindberg and Westman 1990), for example, does not mention Joseph Needham nor his work. Even if

Needham or his work is discussed, the focus remains on "Needham's famous question: What, if anything, can we learn about the causes of the Scientific Revolution by considering the 'failure' of modern science to emerge in any of the other great civilizations of the past?" (H. F. Cohen 1994: 16). However, Needham, in spite of critiques of his work, including that of his famous "Grand Question" – "Why did the Scientific Revolution not take place in China?" – has a special place among the historians and the sociologists of science in the non-West. Kapil Raj puts that clearly and succinctly:

> If there is one belief capable of drumming up a large consensus amongst historians of science, it is surely about western European origins of modern science, most commonly situated in the Scientific Revolution . . . It is in this sea of hubris that Joseph Needham's was a rare voice which did more than just assert uniqueness and superiority of the West in the creation of modern science. (Raj 2016: 255)

"For Needham . . . while modern science originated in the West, yet it is culturally universal . . . 'like rivers flowing into the ocean of modern science'" (Raj 2016: 256). Raj criticizes Needham's "Grand Question," which although "formulated . . . almost a century ago," has shaped "comparative approaches to the history of science" (2016: 256). Placing Needham in "the camp of 'civilisationism,'" Raj seeks to rescue Needham's multicultural argument about the birth of modern science from the civilizational approach, which exemplified a divergence theory. Drawing on two historical examples, one from thirteenth-century China and the other something that happened between the eighth and the tenth centuries in the Abbasid Empire, Raj argues: "Science was, has been, and still is, an interconnected activity with many centres of accumulation" (Raj 2016: 275). Raj is not alone in attempting to recuperate Needham's works, even while certain aspects of Needham's approach are abandoned. Florence Hsia and Dagmar Schäfer, in their introduction to a set of articles published in *Isis* under the rubric "Second Look" that focused on "A Second Look at Joseph Needham," write:

> The "modern science" under which Needham organized his life as a politically engaged scientist and historian served him well as an analytical concept that made a particular version of scholarship possible – and attractive – in a world shattered by conflicts. But recognizing the presentism underwriting the Needham question and driving the entire SCC [*Science and Civilization in China*] project is but a first step toward staking out the grounds on which local and global studies of history of science, medicine, and technology are possible for our time. (Hsia and Schäfer 2019: 97)

There can be little dispute that Needham's ecumenical approach provided an important platform for comparative histories of science, technology, and medicine. However, historians of science seem to remain ambivalent about Needham's Eurocentrism and its impact. Kapil Raj, for example, criticizes the "Eurocentric concept of civilization" and the "ahistorical idea of science." He nonetheless continues to deploy science in the singular and does not (completely) abandon the idea of modern science and its Eurocentric historicization. Raj concludes that his historical examples *"hint at a revision of the view of the emergence of modern science and its periodisation*, but also present-day sciences and knowledge in general" (Raj 2016: 275; emphasis added). Moreover, the categories of the East and the West are not disturbed beyond showing circulations across those categories. Raj writes, "there are certain features which all these knowledge-making activities from East and West share and which cannot be reduced to the conventional division between Western and non-Western science" (Raj 2016: 275).

Florence Hsia and Dagmar Schäfer suggest that "[t]he productive challenges to a Eurocentric field that Needham unspooled over the following years succeeded – perhaps too well" (Hsia and Schäfer 2019: 94). They further argue that "[l]ike so many 'Western advisers' before and since, Needham sought 'to change China' for the better – and on political (leftist) and theoretical (Marxist) terms at odds with those which not only Western 'professional' orientalism (the civilizing mission) but also Western studies of the non-West (liberal modernization theory) had principally developed" (Hsia and Schäfer 2019: 96). "And yet," they add, "the conception of science underwriting both the conditions and contours of Needham's historical work left significant features of such intellectual formations intact" – in particular the "presumption that 'universally' valid world science, that is to say, 'modern science,' had originated in a particular time and place: early modern Europe" (Hsia and Schäfer 2019: 97).

Hsia and Schäfer draw on Dipesh Chakrabarty's *Provincializing Europe* (Chakrabarty 2000) as well, but give Chakrabarty's concern a provincial twist: "Much as the historian Dipesh Chakrabarty has *provincialized Europe in relation to India*, the decolonization of East Asian HSTM [history of science, technology, and medicine], too, has influenced research in meaningful ways" (Hsia and Schäfer 2019: 97; emphasis added). Hsia and Schäfer, in the articulation of their concern with "decolonization of East Asian HSTM," thus end up provincializing Chakrabarty's concept itself. Chakrabarty, as I have reiterated several times in this book, was *not* trying to provincialize Europe geographically (e.g., "in relation to India"). His concern is the Eurocentric historicism that relegates the non-West to a "waiting room of history" (Chakrabarty 2000). Hsia and Shäfer resuscitate Needham's ecumenism by ignoring the Eurocentric historicism that undergirds it – the goal, as they further argue, is to move beyond the "East/West dichotomy" for which "[d]iversity is key, but similarities and convergences" also matter (Hsia and Schäfer 2019: 97).

Indeed, Needham through his ecumenical approach, as many historians of non-western sciences have pointed out, presented a very different idea of the global history of science than those accepted at the high table of the history of science in the West. Needham also directly contributed to the institutional development of the history of science, not just in China, which, so to say, remained his first love, but also in the other non-western countries, such as India (see, e.g., Raina 2020). However, while acknowledging Needham's enormous contributions, we should not forget to carefully situate his work in relation to European imperialism and Eurocentrism. To selectively add or remove different facets of Needham's contributions to suggest that Needham overcame his Eurocentrism or that the Needhamian approach can be extracted from his Eurocentrism would mimic the strategy that is used to defend Eurocentric scholars/scholarship and, more broadly, the impact of European colonialism and Eurocentrism. Moreover, dualist categories of East/West, colonizer/colonized, etc., cannot be simply wished away. We would have to carefully engage with the enduring, albeit complex and multilayered, legacies of these categories in constituting even the history of the present.

In the case of Needham, too, we need to follow Priya Satia's suggestion – "how certain intellectual resources, especially a certain kind of historical sensibility, allowed and continued to allow many people to avoid perceiving . . . inconsistent actions" (Satia 2020: 6).[16] Such a position should not, as Fa-ti Fan's criticism of the post/decolonial theories reminds us, result in reinforcing the dualist divides between the West and the non-West and that the colonizer and the colonized (Fan 2007a, 2007b, 2016). The choice, however, need not be that either we engage with colonialism or we map the complex geographies of science, technology, and medicine (STM). We can map the complex geographies of STM while tracing their colonial genealogies. Colonial encounters, after all, were perceived and impacted differently at different sites (although there were various connections and overlaps in the tactics and practices), and the perceptions and impact of colonialism (and with it that of science) shifted throughout history, depending upon the local and the global contexts. Moreover, we have to remember that there was not a single and unchanging Orient that undergirded the concept and practice of Orientalism (see Inden 1986).

Tong Lam, for example, shows that in China "the desire to overcome the factual deficiency alleged by the colonial powers" became "a major force behind the Chinese intellectual elites' embrace of empirical facts and claims of scientific truth" "in the mid-nineteenth century" (Lam 2011: 7). On the other hand, in Japan, which itself became a colonial power, the relationship with colonialism and science was different. "In Meiji Japan," Hiromi Mizuno argues, "modern science and imperial mythology did not ... appear incompatible with each other because they were neatly dichotomized as 'western science and technology' and 'Eastern ethics'" (Mizuno 2009: 3). However, "the discourse of science after the 1910s" became "very different from the previous discourses ... largely due to the level of industrialization Japan had achieved during and after World War I" (Mizuno 2009: 12). In the context of India, as Shruti Kapila argues, by "the end of the nineteenth century science became the mode of enchantment for an Indian modernity without banishing God," not because India was presented as "'spiritual' rather than rationalist, but

... because religion itself became disenchanted" (Kapila 2010: 131). The complex genealogies of the European colonialism can be seen even in present-day events. As I showed in the Introduction, the United States-versus-China "technological war" cannot be properly understood without reference to the European colonialism, particularly the Opium Wars. Let's also not forget the Orientalist and colonial discursive framing of, for example, Needham's contributions that continues to this day, as I showed at the beginning of this chapter.

Needham, unlike Butterfield, was appreciative of the technological achievements of non-western societies, particularly those of China, and framed the global history of science and technology differently. In the well-known essay, "Poverty and Triumphs of the Chinese Scientific Tradition," which was later published as a part of the book *The Grand Titration: Science and Society in East and West* (Needham 1969a), he historically situated various technological developments in China. Needham provided many examples of techno-scientific developments in China and other non-western societies. "The development of iron and steel technology in China," Needham wrote, "constitutes a veritable epic, with the mastery of iron casting some fifteen centuries before its achievement in Europe" (Needham 1969a: 18). Thereafter, he added, "Contrary to the usual ideas, mechanical clockwork began not in early Renaissance Europe but in Thang China, in spite of the highly agrarian character of East Asian civilization" (1969a: 18). In the field of medicine, Needham argued that "the Chinese were free from the prejudice against mineral remedies which was so striking in the West; they needed no Paracelsus to awaken them from their Galenical slumbers for in these they had never participated" (1969a: 20).

He also corrected various wrongful western appropriations of technological inventions. He thus chided "a writer in the *Times*, no less than the Keeper of the Oriental Books and MSS, in the British Museum itself," who "stated in 1952, with regard to the Tunhuang MSS, that while block-printing was known and used in eighth-century China, it was left to Europeans to devise printing with movable types." Needham bluntly stated: "This is of course nonsense; the second invention is due to Pi Sheng (*fl. c.* AD 1060)" (Needham 1969b: 18). In relation to European claims for the discovery of non-European lands, he wrote: "Europe boasts of the exploratory voyages of Columbus and other navigators. Europe does not so readily inquire into the inventions which made them possible – the magnetic compass and the sternpost rudder from China, the multiple masts from India and Indonesia, the mizen lateen sail from the mariners of Islam" (Needham 1969b: 20).

Needham, thus, did not present European voyages to other parts of the world as discoveries, and he also challenged some of the European claims to such discoveries. For example, he wrote, "Europeans generally speak as if the whole world had been discovered by Europe. This is a very limited conception, not true at all before the Renaissance. The Greeks in Bactria did not discover China; on the contrary, it was the Chinese . . . who discovered them" (Needham 1969b: 20). In fact, Needham even considered China to be a more rational society before the emergence of modern science in Europe. He wrote: "The built-in instability of European society must therefore be contrasted with a homeostatic equilibrium in China, the product I believe of a society fundamentally more rational" (Needham 1969a: 214).

More broadly, Needham emphasized ecumenism of science. Scientific knowledge and technologies from different parts of the world, for him, were like waters from rivers that flowed into the sea of modern science. It has been argued that "Needhamian ecumenism required that scientists bring back culture, society and history into their reflections on science," and as such "Needhamian insight into investigations of history is centered around the key proposition that the history of science is to be interpreted as a history of transmissions" (Raina 1999: 13–14). Dhruv Raina, from whose introduction to the edited volume *Situating the History of Science: Dialogues with Joseph Needham* (Habib and Raina 1999) I quoted above, argues in another article: "Having been liberated from religious opium [because of his Marxist beliefs], he [Needham] found it necessary to extricate himself from the clutches of the Eurocentrism underlying the positivist conception of science" (Raina 1995: 1906).[17]

Needham's ecumenical and multicultural approach to understanding modern science resonated with academics as well as political leaders of the newly independent nations. Needham, for example, categorically stated, "I cannot agree that in the modern world there is such a thing as Muslim science, Japanese science, American or Russian science except as an abuse of the word science, unless one means Muslim, Japanese, American or Russian use of modern science" (Needham 2004: 201). Jawahar Lal Nehru, the first prime minister of independent India, had made a similar declaration in 1948: "Science and technology know no frontiers. Nobody talks or ought to talk about English science, French science, American science, Chinese science. Science is something bigger than countries" (as quoted in Arnold 2013: 363).[18]

For both Needham and Nehru, there did not seem to be any conflict between the Eurocentric framing of modern science and modern science's universality. Nehru, for example, "remarked in 1936 in his autobiography: 'To the British we must be grateful for one splendid gift of which they were the bearers, the gift of science' . . . Without this 'great gift,' India was 'doomed to decay'" (Arnold 2013: 362). David Arnold explains the above-quoted position of Nehru thus: "Nehru could not overlook the authority science had formally enjoyed under colonialism, even while he remained critical of its failure to deliver the practical benefits a backward and impoverished India so patently needed" (Arnold 2013: 362).

Nehru's vision of science and its role in India highlights that he, and through him the postcolonial Indian nation-state, could not escape the Eurocentric and colonial imaginary of modern science that presumes that sciences and technologies could "diffuse" only from Europe/the West and that one should be thankful to European colonialism for the gift of modern science (as though it could not diffuse without colonialism). Nehru could not also see that the above-mentioned assumption was inherently tied to the idea of a singular and universal modern science. This is not to dismiss the idea of modern science that Nehru and other postcolonial leaders and academics have held. Rather, we have to acknowledge that the "legacy of liberal empire [is] so deeply embedded in political and institutional structures and practices that it is difficult for postcolonial societies to shake off; indeed, it is what makes postcolonial societies *postcolonial*" (Satia 2020: 264).

Needham, similarly, could not escape the Eurocentric and colonial imaginary of modern science. He was clear in his belief about the universality of modern science and its origin in Europe/the West. He remained steadfast in this belief and thus could not "extricate himself from the clutches of Eurocentrism." Needham presented the European origin of science, while emphasizing the ecumenism of science: "I disagree with those who call modern science 'Western,' for *though it began in Western Europe it has long ceased to be exclusively Western. I have long pictured modern science as being like an ocean into which the rivers from all the world's civilizations have poured their waters*" (Needham 2004: 201; emphasis added). In fact, even the ecumenism of modern science was shown by him as European in origin. Needham wrote, "Until it [modern science] had been universalized by its fusion with mathematics [with the Galilean 'break-through [that] *occurred only in the West*'] natural science could not be the common property of all mankind" (Needham 1969a: 15; emphasis added).[19]

Hence, according to Needham, prior to this transformation that occurred in Europe, sciences remained "closely tied to the ethnic environments in which they had arisen." As a result, "it was difficult, if not impossible, for the people of those different cultures to find any common basis of discourse" (1969a: 15).[20] Needham's position reflects that "[t]he cultural hold of a certain understanding of history and historical agency was not innocent but designedly complicit in the making of empire" (Satia 2020: 6–7), and such complicity continues in postcolonial times. Indeed, as I have argued in this chapter, Needham's version of the birth of modern science and its diffusion complemented the historiographic model presented by Butterfield and, because of his emphasis on ecumenism in the diffusion of science (before and after modern science was born in Europe), his ecumenical approach played a very important role in solidifying the hold of Eurocentric historicism globally.

From Isis to Clio: unity of science, colonialism, and Orientalism

... our devotion to microstudy has made it harder than ever to engage with our muse, Clio, on a grander scale and in broadly interpretive ways.

It behooves us, then, to think historiographically. Join us in the dialogue between Clio and science.

Robert Kohler and Kathryn Olesko, "Introduction: Clio Meets Science," *Osiris*, 2012: 2, 16

> The subconscious choice and immediate fixation of an
> oriental name requires an explanation.
>   George Sarton, "Why Isis?," *Isis*, 1953: 235.

In 1984, *Isis* published a special "Sarton Centennial" issue dedicated to "vignettes of Sarton, science, and history" (Thackray 1984: 7). Arnold Thackray, the then editor of the journal, reminded the readers: "This issue of *Isis* . . . celebrates not one but two anniversaries connected with 'our' George, George Sarton. In 1884, when Sarton was born, the history of science was entirely an occasional and episodic affair" and "[i]n 1924, when the History of Science Society was launched 'to do something for Sarton in the matter of Isis,' there was more promise than substance to the idea of the history of science as an organized field of learning" (1984: 9). "Today in 1984," Thackray added, "that field is small but secure" (1984: 9). The role of Sarton in building the discipline of the history of science, as Thackray's editorial shows, cannot be overestimated.

As many historians of science have pointed out, the field has moved away from Sarton's positivist and grandiose humanist (science uniting humankind) concerns. However, it is worth revisiting Sarton again, as I have done in this chapter, to explore how his intellectual as well as his professional (in building the discipline) contributions have been a part of a Euro-/ West-centric and Orientalist genealogy of the history of science, which does not normally receive attention (see Elshakry 2010 for a brief discussion). Specifically, revisiting Sarton's Orientalist entanglements would be helpful in understanding how in the decades that followed after "the history of science . . . started off by asking if science was the specific product of Western civilization" (Elshakry 2010: 99), the discipline, very much as I have shown in this chapter with regard to the historical trajectory of modern science, became western and universal (i.e. transcending any location) at the same time.

Sarton did not hide his Orientalist proclivities and entanglements. Of course, we have to remember that during the time period when Sarton was writing and tirelessly working to establish the history of science as a discipline Orientalism did not have negative connotations. In an article published in 1953, Sarton recounted his personal journey that led him to name the history of science journal that he had founded *Isis* (Sarton 1953). He remembered that probably "the name was first engraved in my mind when I read *The Burden of Isis* in 1910" (1953: 237). Nevertheless, as the quote in the epigraph to this section highlights, he was aware that "the subconscious choice . . . of an Oriental name requires an explanation" (1953: 235). He candidly admitted, "The intuitive choice of the name *Isis* betrays the oriental tendencies which were dormant in me" (1953: 239). Interest in "the Orient" had been inculcated in Sarton in his "early childhood," when he came across and read "a French account of Chinese philosophy" that was available in his "father's library." This "orientalism was unconscious as such" and Sarton saw it as "simply an international and human interest" (1953: 239).

Sarton's autobiographical history of the birth of *Isis* nicely illustrates Edward Said's exposition of Orientalism as a "result of cultural hegemony at work" (Said 1979: 7). In Sarton's account, too, "the Orient" is discovered, but the discovery, which consists of the representation of the Orient as the "other," is constrained by the existing Orientalist cultural representations. Moreover, this discovery is staged for the West and serves some function in the context of the West. To clarify, it is not that the Orientalists were, or are, (necessarily) negatively oriented towards the non-West. Indeed, much of the European colonial ideologies and practices, most ostensibly the aspects that were expressed through the civilizing missions, were aimed at taking care of the "other," and many Orientalists, for example William Jones, were guided by their love for the early historical heritage of the colonized non-West. The unwavering characteristic of Orientalism has been, and continues to be, the constitution of the non-West as the "other" that is "revealed" to the West to serve the purposes of the West, wherein the westerner (intellectual, policymaker, etc.) represents the "other" and speaks for him/her/them.

Sarton lamented that he had "grown to hate" "political history" because it consisted of "European history focused on Belgium" – "no American history at all, except mention of Columbus" and "little of Oriental history" (Sarton 1953: 239). His interest in the Orient thus emerged from the cultural inheritance. "Ancient *Egypt was revealed*" to Sarton "in 1905, when . . . [he] visited the Egyptian section of the museum of antiquities" (1953: 235; emphasis added). He also had other cultural encounters with the Orient. For example, a visit to a Japanese art exhibition in London in 1910 became his "first *revelation of Japanese art* outside of wood engravings" (1953: emphasis added). He also came to know about "Hindu music" and "Hindu and Indonesian art" during this period. He admitted that he was "unable to master the oriental languages but . . . *tried to understand the soul of Asia* as expressed by her great artists"(1953: 236; emphasis added).

Sarton was torn between "the history of science and the history of oriental arts" as the choice for his intellectual pursuit. He recounted: "I was like a man who falls in love with a strange woman, yet remembers in time that he is already married. I had to reach a decision, a very hard one. Adieu *Kuan Yin*!" (1953: 236). Nevertheless, as he added in his final recollections, "my internationalism and my orientalism grew apace together, being the natural fruits of my interest in . . .

the whole world, the whole mankind" (1953: 239). The choice of Isis – an Egyptian goddess – as the name of a history of science journal was not clear to Sarton when he established the journal in 1913, but as he "learned to understand . . . the history of science should include Eastern as well as Western science, the title was more completely justified" (1953: 241). Nonetheless, it was "impossible . . . to devote much space in *Isis* to Eastern science . . . even if her [Eastern science] capacity was considerably greater than it is" because the "development of modern science . . . is so important and of such exuberance in every direction that the briefest account of it requires considerable space" (1953: 241–2). In the emergence of Sarton as a historian of science and that of the journal that he founded, the Orient as the "other" of Europe/the West remained integral to the identity of not only the western "self," but also of modern science and its international scope defined through the unity of science, which became the focus of the history of science.

Almost six decades after Sarton's article on the history of naming *Isis*, in 2012, when *Osiris* published a special issue whose goal, according to the editors Robert Kohler and Kathryn Olesko, was "less consciousness raising than consolidation" (Kohler and Olesko 2012: 6), it was titled "Clio Meets Science: The Challenges of History." Interestingly, Sarton, in his account of how the name "Isis" was chosen, had also briefly discussed the naming of *Osiris*: "When the need of a new journal arose to take charge of the longer memoirs, its name was easy to find, *Osiris* . . . Every reader understood without explanation that Osiris and Isis, brother and sister, man and wife, represented two undertakings as closely related as could be" (Sarton 1953: 238). In the *Osiris* special issue, the representational trope used to define the "self" of the history of science had evidently shifted from the Orient to the Occident. Moreover, the Orientalist and colonial entanglements in the birth of

the history of science and, more broadly, in the historiographic framing of the unity of modern science
was almost completely ignored. The editors' introduction and the three essays under the title "Reflections"
focused on and problematized the unity of science
and the disciplinary identity of the history of science
in which the non-West became one of the contexts in
which the unity of science was troubled: Additionally,
the non-West was also the focus of a couple of explorations of the "horizons" of the history of science.

The editors, Kohler and Olesko, as the quote in the
epigraph of this section shows, exhort the readers to
"think historiographically" and "[j]oin ... in the dialogue between Clio and science" (Kohler and Olesko
2012: 16). Indeed, as a simple Google search would
show, Clio – the muse of history in Greek mythology
– is a popular signifier in the western imaginary that is
used as the name of a number of things and activities,
ranging from legal software, the name of towns and
clubs, to the name of a brand of snacks. However, when
it is deployed as a call to "think historiographically," it
would also be useful to remember that the name Clio
"in Greek means to celebrate or glorify" which was
"the earliest purpose of historical writing," and Clio's
"statues in the Vatican, the Louvre, and elsewhere usually represent this primate of the Muses" (Munro 1941:
403).

I am certainly not suggesting that the essays in
this special issue were attempts at glorifying the past.
Instead, I want to highlight two issues. First, a reference
to a term such as Clio, without a careful engagement
with the term's discursive role in the western imaginary,
appears to constitute the history of science as western,
even when it seeks to present the discipline as transcending any locational identity. Unfortunately, some
of the essays in this special issue echo such an identity
of the history of science. For example, Paul Forman's

essay (Forman 2012) on the impact of modernity and postmodernity on the discipline, ensconces the debate exclusively within the West (it does not even mention the terms non-West, colonialism, etc.) and seeks to give an even deeper western identity to science as well as to the history of science, albeit presented as characteristics of modernity or postmodernity. Second, ignoring the role of the non-western "other" in the history of the history of science and in the imagining of the unity and universality of modern science results in a partial picture that would not allow us to escape the lingering effects of European colonialism and Orientalism.

Indeed, as Jan Golinski, in his "Reflections on Singular Science and Its History" in the same special issue of *Osiris*, states: "To speak of science as a single thing suggests a degree of unity, exclusivity, and long-term continuity that the historical record does not seem to manifest" (Golinski 2012: 19). Nevertheless, he warns us against too easily dismissing the lingering impact of the belief in a singular science. Golinski highlights the reasons behind the difficulty in accepting "science as a single thing," among which he lists the difficulty in locating science "in its familiar form" "when non-Western cultures are brought into focus" (2012: 19). He does not, however, discuss the entanglements of the belief in a singular science with colonialism and/or Euro-/West centrism. Peter Dear in his essay, which is a part of the "Reflections," critically engages with Euro-/West centrism. Dear, for example, writes: "Although Latour's explicit intent was to dissolve the 'Great Divide' between the West and the rest of the world, the net effect of his work has been to reinforce the view that science and its associated physical power was exported wholesale as a fundamentally European commodity" (Dear 2012: 47).

Dear reminds us that "[t]here is nothing novel in proclaiming the diversity of those activities and bodies of knowledge that we call *science*" (2012: 37). Rather, "[t]he reality of what science truly is resides less in essential similarities between its assorted branches than it does in the ideology that binds them" (2012: 38–9). Moreover, "[o]nly the presence of this ideology can identify 'science' in its modern sense: not particular practices, or specific ideas, but a self-effacing ideological construct that makes claims going beyond what it can fully deliver" (2012: 55). "The history of this ideology, made up of cultural, intellectual and social reifications," Dear argues, "is a coherent project for writing the history of modern science" (Dear 2012: 55).

The only thing I would like to add, as I have shown in this chapter, is that this ideology of modern science was, and continues to be (although much less explicitly visible as such), entangled with European colonialism and Orientalism, and if we do not carefully engage with the latter, we will continue to replicate colonial and Orientalist tropes, despite our expressed intention to move beyond those tropes (see Abraham 2006 and chapter 3 for a detailed discussion). Simply showing science as multiple and situated would not allow us to transcend the enduring legacies of the European colonialism. As I show in the next chapter, genealogies of European colonialism and Eurocentric historicism continue to haunt, for example, some of the science and technology studies (STS) when they engage with the non-West and/or colonialism and seek to move beyond Eurocentrism.

3

Colonialism and Euro-/ West Centrism: Postcolonial Desires, Colonial Entrapments

> Here . . . is a kind of question, let us still call it historical, whose *conception*, *formation*, *gestation*, and *labor* we are only catching a glimpse of today.
> Jacques Derrida, *Writing and Difference*, 1978: 293

"Where is East Asia in STS?" (Lin and Law 2019). "Provincializing STS: Postcoloniality, Symmetry, and Method" (Law and Lin 2017). "We Have Never Been Latecomers!? Making Knowledge Spaces for East Asian Technosocial Practices" (Lin and Law 2015). These are the titles of a few essays from a series of recent publications by Wen-yuan Lin and John Law, the latter being one of the founding formulators of actor–network theory (ANT). In these articles, Lin and Law have presented a new manifesto for not just postcolonial science studies but also for the broader field of science and technology studies (STS). Law and Lin state that "STS is not short of studies on postcoloniality" (Law and Lin 2017: 211) "At the same time," they highlight, "the discipline has usually made use of Euro-American analytical terms" (2017: 212).[1] Their goal, is to "argue for *forms of postcolonial investigation that use non-Western analytical resources*" (2017: 212; emphasis added). The analytical move that they are presenting is, however, aimed not only at postcolonial studies and postcolonial

111

or non-western contexts. "Our main concern," Law and Lin declare, "is thus to ask what might happen if STS were to make more systematic use of non-Western ideas" (2017: 212).

The concern for Law and Lin, through their proposed new agenda for STS, is also to disrupt what they call the "analytical-institutional complex" of STS that "works to reproduce postcolonial intellectual asymmetries" (Law and Lin 2017: 213). They acknowledge the danger of creating "a version of exoticizing Orientalism, that knowledge from outside Euro-America offers special or privileged access to reality." And they also do not wish "to get caught up in chauvinistic 'national science' projects" (2017: 214). They argue for what they call a "third postcolonial version of the principle of symmetry" that would make the traffic of "scientific travel" "lively, two-way, and contested." They add emphatically, "because there is no single postcoloniality, there would be multiple centers, a variety of postcolonial symmetries, and a series of different STSs," and as such they ask us to imagine "scientific travel" through "translation and its betrayals" (Law and Lin 2017: 214; Lin and Law 2014).

In this chapter, and indeed the whole book, my goal is to map the genealogies of European colonialism and their constitutive role in the *history of the present*. I do not offer an alternative model for postcolonial science studies or for that matter for STS. I have deliberately chosen this path, in part, because, European colonialism, I think, continues to occupy a strange place within STS. Even when it is criticized, the violence (physical, epistemic, ontological, etc.) of colonialism and its enduring impact often get quickly passed over or are erased (Prasad 2017). I am certainly not suggesting this is true of all STS. Such a tendency is, however, evident in a dominant section of STS. And the fact that, instead of critically engaging with this tendency, it is ignored and even hailed as *postcolonial* intervention says a lot about the strange place of colonialism within STS.

My concern is twofold. On the one hand, there are empirical elisions of the history of colonialism and its impact in some STS work that result in presenting West-centric stories, even when that work is exclusively discussing an erstwhile colonial context. On the other hand, there is a failure "to see Others ... as historically constituted" and a tendency to treat the "other" as an "ontological given" (Said 1989: 225). Law and Lin, for example, express an important concern with regard to lack of engagement with the non-western scholars and how that is reflective of an "analytical-institutional complex." A similar concern is raised by Dipesh Chakrabarty, whose concept Law and Lin draw upon, but unlike Chakrabarty, Law and Lin, as I show in the following, end up constituting the "other," including Chakrabarty, as an ontological given and forget that the "other" is historically constituted. More broadly, to appropriate Achille Mbembe's claim with regard to his critique of "black reason," we need to be mindful that Euro-/West centrism "is a complicated work of doubling, uncertainty, and equivocation, built with" colonialism/coloniality "as its chassis" (Mbembe 2017: 10).[2]

Law and Lin's article "Provincializing STS" (Law and Lin 2017), for example, draws on the concept of provincializing from Dipesh Chakrabarty's widely cited book *Provincializing Europe* (Chakrabarty 2000).[3] However, they give a twist to Chakrabarty's argument. Law and Lin write:

In his *Provincializing Europe* (2000), he [Chakrabarty] describes how fifteen thousand tribal people were massacred in Bengal by the British in 1856. After the first deaths the victims kept on coming. But why? The survivors said that their God had told them to fight. He would protect them. The story is brutal and horrific, but Chakrabarty's particular concern is analytical. *As a Western-trained historian he knows that gods are not really powerful. But as an Indian this makes him deeply uneasy. Here, then, is the question: To which should he give priority, Western historiographic convention or a world in which gods (not just beliefs about gods) cause actions?* (Law and Lin 2017: 214; emphasis added)

They argue that "[t]he principle of symmetry" – between true and false beliefs (Bloor 1991) and between humans and non-humans (Callon 1986; Latour 1987), which have been the guiding principles for STS – "catches a part of" what they describe through Chakrabarty's book (Law and Lin 2017: 213). Hence, they suggest, "it is time to extend it again" (2017: 214). They thus propose a "postcolonial version of symmetry" that allows us "to think about translation and its betrayals" (2017: 214). Law and Lin cite Casper Bruun Jensen and Anders Blok's (Jensen and Blok 2013) and Atsuro Morita's (Morita 2014) studies in this regard. Interestingly, Jensen and Blok's goal seems somewhat different – at least from the expressed objective of Law and Lin in "provincializing" STS.

Jensen and Blok explicitly state: "Rather, than adopting a strategy of 'provincializing Europe' (Chakrabarty 2000) . . . we view Japanese techno-animism as a tool with which to potentially *exoticize* European naturalist practices more thoroughly than even Latour has so far been capable of" (Jensen and Blok 2013: 109; emphasis in the original). (We are all exotic now!). For them, "the issue is to bypass the twin perils of Orientalism and Occidentalism *in the same move*" (2013: 109). I must admit I am left almost speechless reading such a claim. But let me nevertheless add: Orientalism and Occidentalism are not simply methodological tricks to be avoided in our academic research and writing. They have been, and continue to be, powerful discourses that have constituted and continue to constitute selves and practices long after formal European colonialism ended, as has been pointed out by many scholars, including more recently by Achille Mbembe in his *Critique of Black Reason* (Mbembe 2017). I specifically cite Mbembe because he also discusses the return of animism, though his concern is radically different.[4]

Atsuro Morita, who is the other scholar cited by Law and Lin in support of their proposed postcolonial symmetry, in fact alerts us to the need for self-reflexivity. While highlighting the Latourian concern that "following ontological experimentations done in laboratories is far more interesting than posing a somehow conventional question of reflexivity," he writes, "[a]lthough I am sympathetic to ANT's ambition, its strong emphasis on following the actors seems to *relegate the problem of the researcher's own analytical device to a blind spot*" (Morita 2014: 230; emphasis added). In a significant way, Law and Lin's postcolonial studies reflect a relegation "of a researcher's own analytical device to a blind spot." Since my concern is to map the genealogies of colonialism, and this chapter focuses on tracing those genealogies within the field of STS, let me further probe the blind spot that Morita highlights.

Let's look at the discursive framing of colonialism and colonial violence in the quote that I presented earlier, in which Law and Lin draw on Dipesh Chakrabarty's work to propose their own model of postcolonialism and of provincializing STS. To start with, Law and Lin frame Chakrabarty's intellectual concern (and his biography) through a West/non-West (Indian) dualist frame. Chakrabarty is presented as a "western-trained historian," but "an Indian." In doing so, Chakrabarty's biography as well as his thought are Orientalized.[5] Interestingly, Chakrabarty's training before he joined the Australian National University for his PhD was in

India (at the Presidency College and the Indian Institute of Management, both in Kolkata).

More importantly, when Chakrabarty states "I too write from within this inheritance" of "European thought" (Chakrabarty 2000: 7), he is, on the one hand, acknowledging the continued intellectual power and reach of European thought and, on the other, distancing himself from a dualist divide such as the one presented by Law and Lin. Chakrabarty laments the consequence of this Eurocentric embeddedness of even non-European social sciences and humanities, writing, "few if any . . . social scientists of India would argue seriously with, say, the thirteenth-century logician Ganesa or with the grammarian and linguistic philosopher Bartrihari (fifth to sixth centuries), or with the tenth- or eleventh-century aesthetician Abhinavagupta" (2000: 5). This outcome, though, as he explains, is itself an effect of Eurocentric historicism that "posited historical time as a measure of cultural distance . . . between the West and the non-West" (2000: 7). There is thus no *tabula rasa* that can allow us to simply bypass the impact of European colonialism.

In Law and Lin's appropriation of Chakrabarty's argument, the physical violence of colonialism is mentioned – "[t]he story is brutal and horrific" – but then quickly passed over. Chakrabarty's analytical concern, however, is not erasure (or reinsertion) of gods per se, but what their insertion into historical writing allows us to show, namely, historical actions that are outside the Eurocentric historicism that constitutes the non-European/western world as a "waiting room of history" through the temporal order of "first in Europe, then elsewhere" (Chakrabarty 2000). Colonialism and its impact and Chakrabarty and his analytical intervention, are, thus, appropriated (discussed but quickly ignored) so as to present the pluriverse of actor–network theory (ANT). Ironically, in the process, Law and Lin seem to abandon one of the most cherished contributions of ANT, namely, hybrids and how hybrids transgress and transcend dualism, which, as Latour (1993) insisted, is a modernist fiction. In their effort to reinsert the non-West through an extension of the STS principle of symmetry, Law and Lin, even though their concern with the analytical/academic exclusion of the non-West is right, reinforce the philosophical basis of dualism – that the two (or more) constitutive elements of dualism are separate and different.

Let me clarify. My concern is not that Law and Lin's interpretation of Chakrabarty is wrong or that it could be better. I am not interested in any boundary keeping. Rather, Law and Lin's appropriation of Chakrabarty's theoretical intervention in relation to European colonialism and Eurocentric historicism reflects an ethnocentrism that Jacques Derrida called "European hallucination" – an "interested blindness" (Derrida 1998).[6] Law and Lin's postcolonial intervention is more like another version of the now widely discredited multicultural theory, albeit translated in the context of colonialism/globalization. Colonialism, not just in this essay but more broadly within ANT, is presented simply as a mediator between the two worlds of colonial encounters that does not profoundly disrupt or change the encountered worlds. To quote Bruno Latour's observation on another colonial encounter, "Lapérouse was able to put Sakhalin on a map, but the South Pacific cannibals that stopped his travel put him on *their* map!" (Latour 1987: 229).

I agree with Warwick Anderson that "actor–network theory ... is an incredibly powerful analytical tool" (Anderson and Prasad 2017: 136). Steven Shapin rightly complimented ANT, Latour in particular, for opening a "particularly significant line of thinking about scientific travel" that draws "attention to the ways in which patterns of military domination, colonialism and

worldwide trade have established channels which inte-
grate the world and which standardize its knowledge
and its practices" (Shapin 1998: 7).[7] Anderson goes a
step further and suggests that ANT "could be made to
do postcolonial work, as De Laet and Mol, and Law,
have tried to show"; adding, he believes "Law and Lin
are advancing a rapprochement between postcolonial
critique and ANT" (Anderson and Prasad 2017: 136).

I think ANT and, more broadly, STS need to much
more carefully and extensively engage with colonialism and coloniality and explore whether and how that
necessitates a radical retooling of the accepted analytics
and methodologies of STS. The deployments of science
and being scientific as "tools of the empire" to hier-
archically define people and societies (see, e.g., Adas
1989; Headrick 1988), which continued well into the
twentieth century and whose genealogical links can be
seen even now, would not wash away just by showing
that science is multiple and situated. Analogically, that
would be like believing racism and sexism would go
away after the falsity of their biological (and as such
ahistorical) basis has been shown and their situated
articulations highlighted.

In this chapter, I point out genealogies of colonialism in some ANT studies and contradictions in Sandra
Harding's and Warwick Anderson's engagements with
colonialism and the non-West not because I do not
appreciate ANT's, Harding's, or Anderson's contributions in offering new tools to study different facets of
science, technology and society. My critique of certain
aspects of ANT, for example, does not stop me from
productively engaging with many ANT insights (in
fact, I was taught that critique itself was a productive engagement). I also do not wish to suggest, as this
chapter should make clear, that such a problem exists
only with ANT in the field of STS. My concern is much
broader and also urgent, and that is the other reason for

my focus on mapping the genealogies of colonialism to excavate the history of the present. As Priya Satia, drawing on Stuart Hall, writes "the national [or civilizational] frames of old historical narratives cannot serve us; colonial history has made it impossible to conceive of specific communities and traditions with settled and fixed boundaries and identities" (Satia 2020: 265). If STS continues to engage with histories and sociologies of science and technology without carefully and critically engaging with the genealogies of colonialism, it runs the risk of forgetting not only what has happened, and continues to happen "out there," but also what is happening in its own Euro-American-Australian backyard. Satia bluntly reminds us of the dangers of such forgetting: "The Brexit movement . . . is a sort of memorial in the antithetical direction, the very opposite of coming to terms with the imperial past in favor of instead glorifying its memory" (Satia 2020: 275). Let's also not forget the attempted takeover by force of the United States Congress by white supremacists on January 6, 2021 because they did not believe that the presidential election was fair – the state of exception that defined the American white-settler colonialism is central to such white supremacist claims and actions.

A lack of critical engagement with colonialism and its enduring impact would also make international collaborations on urgent matters of concern such as climate change very difficult. For example, use of the term "Anthropocene" has become ubiquitous and is increasingly being criticized for ignoring the entangled "histories of race and racism, capitalism, and European imperialism." As a result, there have been calls to deploy other terms such as "racial capitalocene" (Verges 2017). Nicholas Mirzoeff similarly declares that "it's not the Anthropocene, it's the white supremacy scene" (Mirzoeff 2018). Moreover, as I showed in chapter 1, genealogies of colonialism can be seen even in the ways misinformation and conspiracies were being framed and interpreted during the COVID-19 pandemic.

In the next section, I critically engage with two well-regarded and highly cited articles in the field of STS, namely Marianne de Laet and Annemarie Mol's "The Zimbabwe Bush Pump: Mechanics of a Fluid Technology" (Laet and Mol 2000) and John Law's "On the Methods of Long-Distance Control: Vessels, Navigation and the Portuguese Route to India" (Law 1986). Both these articles were published in very reputable journals, the first one in the *Social Studies of Science* and the second in the *Sociological Review*.

In the "Parliament of Things," whose voices/ inscriptions are getting recognized?

Thus he ... forms an interesting subject for an ANT-type analysis, because the ANT-tradition rarely works by adding to what has already been established. Instead it introduces variations, sets up contrasts, and, time and again, proposes shifts. ... The story of the Zimbabwe Bush pump suggests that an actor is not necessarily a hero who designs the strategy of the army ... Rather than taking control, actors may also seek to serve the world around them.

Annemarie Mol, "Actor–Network Theory: Sensitive Terms and Enduring Tensions," 2010: 256

The "he" in the quote above is Peter Morgan, the "non-classical hero" of Marianne de Laet and Annemarie Mol's article "The Zimbabwe Bush Pump" (Laet and Mol 2000). Laet and Mol do not hide their love and awe for Peter Morgan. They write: "Serving the people, abandoning control, listening to *ngangas*, going out to watch and see what has happened to your pump: this is not the line taken by a sovereign master." "Here we have, instead, a *feminist dream of an ideal man*" (2000: 252; emphasis added). They clarify in a footnote that they "do not mean to say anything about *Peter* Morgan, in a personal sense," because they "cannot claim to know him 'personally' . . . With 'ideal man,' here, we refer to Dr Morgan, a public figure in the space of technology design and water politics" (2000: note 58; emphasis in the original). In another article, Annemarie Mol clarifies the use of the term "feminist." Mol writes, "[f]eminism in ANT is not a matter of repeating the categories 'man' and 'woman' in order to see oppression at work everywhere, but of shifting and changing them" (Mol 2010: 256, note 14).

Nevertheless, Peter Morgan is an ideal man, as Laet and Mol reiterate, also because he is "a non-classical hero who is as active as can be, and yet who makes no claims to *heroic* authorship." Much like the Zimbabwe bush pump, "he too is fluid, *dissolving* into his surroundings" (Laet and Mol 2000: 226). Fluidity, as Laet and Mol state at the outset, is "the specific quality that attracts us to the Zimbabwe Bush Pump" (2000: 225). The "Bush Pump is not a solid character." And as such the pump highlights that "[n]ot only can actors be non-rational and non-human; they can also . . . be fluid without losing their agency" (2000: 227). The fluidity of the bush pump is, however, "not an accident. The Bush Pump is made that way. *It is made that way by a modest inventor*" (2000: 225; emphasis added). Although, as the title of Laet and Mol's article suggests, the Zimbabwe bush pump is the actor whose fluidity is in focus, Peter Morgan is the man who makes the *becoming* of that non-human actor possible.

The bush pump, Laet and Mol argue, "is not well-bounded but entangled, in terms of both its performance and its nature, in a variety of worlds" (2000: 227). "The Zimbabwe Bush Pump has existed for more than half a century, but it has not remained the same." They explore the socio-technical boundaries of the bush pump genealogically and also in the present. "The current model [i.e., Type B] results from restyling and improving an older manually operated water pump that was first designed in 1933 by Tommy Murgatroyd in what was then Rhodesia's Matabeleland." And the "experimenting and changing are still going on" (2000: 228). The entangled boundaries of the bush pump extend even further – "the Zimbabwe Bush Pump is a nation-builder that gains strength with each new installation" (2000: 236). The installations, through which the nation building is also occurring, as Laet and Mol show, is distributed as well. It, for example, involves active community participation – "sometimes operating the rig [for the boring of the water hole] turns into a village feast. Village women push the iron crossbar to drive the auger into the ground, while village men sit on the bar to weigh it down and children dance around" (2000: 233). In the distributed actions surrounding the bush pump, the actors/actants are not only this-worldly; the other-worldly is also included. The "instruction manuals and other publications" "state clearly and repeatedly" "that local water diviners should be consulted before any decision about the siting of a water hole is made" (2000: 234).

Laet and Mol thus present a wonderful (and shall I say heartwarming) story of not only a historically and socially distributed artifact – a fluid technology – but also that of an inventor, Peter Morgan, "the actor behind the Pump, who refuses to act as such." They reiterate their admiration for Peter Morgan in the concluding section of their article: "Dr Morgan's carefully sought dissolution, his deliberate abandonment, is not simply an asset in any man, but is especially suited to the dissemination of the Bush Pump." They add: "Pleased with what he [Morgan] calls the 'forgiving nature' of the Bush Pump, he has made it after his own image – infused it with fluidity that he incorporates himself as well" (2000: 252).

Nonetheless, Laet and Mol "do not want to set up *fluidity* as a new standard." Their concern is "[h]ow to be normative when there is no single, self-evident standpoint to speak from?" They conclude, "by using notions such as *love*, we want to signal how we are interpellated by it. So maybe this is an exercise in praise after all. For we never set out to pass judgement on the Zimbabwe Bush Pump, but have allowed ourselves to be *moved by* it. And this paper is an attempt to move you, reader, too" (Laet and Mol 2000: 253; emphasis in the original).

Let me start where they end. Laet and Mol, as they accept in the above quote, were *interpellated* by values such as love that are embodied in the Zimbabwe bush pump and evident in the actions of Peter Morgan. Interpellation is a term used by the late Marxist scholar Louis Althusser to signify that *"all ideology hails or interpellates concrete individuals as concrete subjects"* (Althusser 1971: 173). Moreover, according to Althusser, "there is no ideology except for concrete subjects, and this destination of ideology is only made possible . . . *by the category of the subject* and its functioning" (Althusser 1971: 170). In short, we are hailed as a subject via ideology and our functioning, which reflects our ideological interpellation and gives expression to that particular ideology. I am not imposing an Althusserian interpretation on Laet and Mol. In the footnote accompanying their use of the phrase "we are interpellated by it," they write: "In good Althusserian fashion, one may doubt whatever one is seduced by" (Laet and Mol 2000: 262). I am glad, whatever the intent of Laet and Mol in invoking the concept of interpellation, they at least agree that we can be interpellated as subjects via ideology or discourse. Does the article's framing thus reflect their interpellation by a particular discourse or ideology? And can a critical engagement with that ideology or discourse help us to perhaps tell the story differently?

Laet and Mol, drawing on Michel Serres, emphasize that "the Bush Pump contains its environment" (Laet and Mol 2000: 262, note 61). But what do Laet and Mol consider as constitutive of the bush pump's environment? Does the colonial context of Zimbabwe have any place in this environment? The colonial name of Zimbabwe, Rhodesia, is mentioned twice in the article, first when Laet and Mol inform the reader that "an older manually operated water pump [to which the Zimbabwe Bush Pump genealogically belongs] . . . was first designed in 1933 by Tommy Murgatroyd in what was then Rhodesia's Matabeleland" (2000: 228), and then while presenting the "displacement" of Dr Peter Morgan from Malawi to Harare on the "invitation by the Zimbabwean (then Rhodesian) Ministry of Health" (2000: 248). They do not tell us that *Dr Morgan was born in England* and grew up in Gidea Park and received "a PhD in marine biology at the University of Hull before moving to Malawi."[8] In 1972, when Dr Morgan moved to Harare to work as a medical research officer at the Blair Research Laboratory, which was renamed the National Institute of Health Research in 2011, Zimbabwe was still under white-settler colonialism.[9] In fact, Zimbabwe was in the midst of the second *Chimurenga* – a bitter war for independence from white-settler colonialism – at that time. It achieved independence in 1980, but only after a biracial settlement defined through the "Lancaster House Constitution" was agreed upon by all parties.

I discuss the colonial context of Zimbabwe because it was, and continues to be, directly a part of the "environment" of the bush pump. It can explain why, for example, water shortage and lack of arable land almost exclusively affects the Black population of Zimbabwe; and why almost all of the photographs and drawings in Peter Morgan's book (Morgan 1990), which Laet and Mol use extensively as the source of their information about the bush pump, are those of the Black people. Ever since 1890, when "Cecil John Rhodes' Pioneer Column hoisted the Union Jack in Salisbury (Harare) . . . and claimed the country for Britain," "forced appropriation of land was the policy of choice of the colonial settlers and their successive governments" (Mlambo 2005: 4). Cecil Rhodes had famously said "I would annex the planets if I could . . . Expansion is everything" (as quoted in Arendt 1945: 441). Rhodes eventually "settled" for colonizing southern Africa, another temperate part of the planet, which Theodore Roosevelt had claimed had to be kept for the whites. This situation resonates with the answer of W. E. B. Du Bois, a contemporary of Rhodes and Roosevelt, to a question that he himself posed in his essay titled "The Souls of White Folks" – "But what on earth is whiteness that one should so desire it?" Du Bois wrote, "I am given to understand whiteness is the ownership of the earth forever and ever, Amen!" (Du Bois 2014 [1920]: 16).

After 1930, when the colonial government passed the Land Apportionment Act, it became "illegal for Black Africans to own land outside the established 'native reserves,'" which "comprised some 22 percent of the country's least arable land, while the white people . . . who constituted 4.5 percent of the country's population, gained access to around 51 percent of the country's most arable land."[10] "At independence in 1980, 6,000 white farmers occupied 15.5 million hectares of land, while millions of Africans remained squashed in the Reserves, now euphemistically renamed Communal Areas" (Mlambo 2005: 7; Moya 2004). Let's also not forget that the "whites remain[ed] in control of most of the arable land in the country" even after independence

(Mlambo 2005: 12). To make matters worse, land reforms, even after independence, were not easy because of the Lancaster House agreement that was facilitated by the British and led to the independence of Zimbabwe. This agreement made it mandatory that the independent Zimbabwean government could acquire land for redistribution only on a "willing buyer, willing seller basis" and the buying of the land (from the descendants of the white colonial settlers) had to be at the market price and in foreign currency (Mlambo 2005).

The Lancaster House agreement would have failed had Britain not agreed to provide financial support for the resettlement, and the British did provide some support in this regard. Nevertheless, because of these constraints, it was observed that after independence, in the 1980s, "over 70 per cent of land acquired for resettlement through the market ... [had] been agro-ecologically marginal and located mainly in the drier, more climatically erratic southern regions of the country" (Mlambo 2005: 12; Moya 2000). Racial inequity and tensions thus continued to rage in post-independence Zimbabwe.

During this period "white Zimbabweans struck a subtle balance between articulations of moral virtue and claims of technocratic expertise." The moral virtue of the white Zimbabweans entailed erasing the racialized colonial history of land appropriation and fashioning their moral "claims to belonging ... derived not from the politically fraught terrain of race and citizenship" (Suzuki 2018: 85). Such a moral position, which, apart from emphasizing generational rootedness, was tied to how "many whites ... affiliated themselves with the land rather than with surrounding societies," thereby presented an "environmentalist form of identity" (Hughes 2006: 269), hiding the racialized nature of the land-related claims of the Zimbabwean whites.

The technocratic expertise of the white Zimbabweans, which arose as a result of "monopoly over expert knowledge that formerly was ensured through systematic racial discrimination in terms of education, technical training, and opportunities for skilled employment" (Suzuki 2018: 83), also made them indispensable in the postcolonial era. This hierarchy based on technocratic expertise was further reinforced because of "the nationalist [post-independence] government's modernist, top-down approach, which staunchly believed in technology transfer" (Mavhunga 2014: 222). And this was not unique to Zimbabwe; postcolonial governments commonly used science and technology as the reason of the state (Prakash 1999). In the context of Zimbabwe, the continued support for Dr Peter Morgan (after his stint at the Blair Institute until 1990, he worked as the scientific advisor to the Ministry of Health during 1990–2), although based on his contributions to public service that are widely acknowledged across racial lines, was also tied to the racialized monopoly of technocratic control.[11]

Moreover, Dr Morgan's oft-repeated reference to the community as the maker of the bush pump is perhaps not simply a reflection of his humility; it also tells us something about post-independence Zimbabwe. As Sister Patricia Walsh of St Theresa's Mission Hospital wrote just a few years after Zimbabwe's independence, "great stress . . . placed on community participation at all levels of decision-making in [newly independent] Zimbabwe" contributed further to the support of Dr Morgan (Walsh 1983: 14). Focus on community participation did not reflect just the initial enthusiasm of a newly independent country. "Independent Zimbabwe," as the World Bank's Water and Sanitation Program report highlighted in 2002, "inherited a highly centralized system of government, and the new government began devolving responsibility and power to lower levels of government soon after independence," and this required a series of ongoing policy measures (Robinson 2002: 4). Moreover, as Clapperton Mavhunga argues, among the indigenous communities of Zimbabwe there is a strong belief that "[k]nowledge individualized was dead. Vedzimbahwe did not write on paper; orality, language, and communality are what brought and kept information alive from generation to generation" (Mavhunga 2017: 59). That is to say, a focus on the role of community in post-independence Zimbabwe highlights the truly distributed nature of the bush pump's development and use.

The colonial genealogy of land grab that resulted in the dispossession of native Africans and the monopoly of white Zimbabweans (ensured through the racialized policies of white-settler colonialism) is thus directly a part of the "environment" of the bush pump. Unfortunately, this colonial genealogy is not even mentioned by Laet and Mol. They passingly mention in a footnote that "[a] 1997 government initiative to relocate land from wealthy farmers to the rural poor has been analyzed in newspaper reports and scholarly commentaries as an effort to strengthen the withering support for Robert Mugabe's ruling coalition in Zimbabwe – and thus, as an exercise in nation building" (Laet and Mol 2000: 260, note 36). Laet and Mol are partially right. Mugabe did indeed initiate what he called the third *Chimurenga* in "an effort to strengthen withering support." However, the racialized appropriation of land and resulting inequity in land ownership that undergirded Mugabe's claim were not false, even though the "arbitrary" enforcement of this new phase of land reforms was problematic and eventually led to Mugabe's own downfall.

The 1997 initiative for land redistribution cannot also be separated from Britain's backtracking from its promised financial support for land reforms in Zimbabwe after the latter's independence. In 1997, when Tony Blair's

New Labour government came to power, Clare Short, then secretary of state for international development, "wrote to the Zimbabwean Minister of Agriculture and Lands, Kumbirai Kangai: 'I should make it clear that *we do not accept that Britain has a special responsibility to meet the costs of land purchase in Zimbabwe.* We are a new Government from diverse backgrounds without links to former colonial interests. *My own origins are Irish and as you know we were colonized not colonizers*'" (as quoted in Mlambo 2005: 14; emphasis added).

This change in British policy occurred when Zimbabwe's economy was doing very badly (in part as a result of the structural adjustments in the economy enforced through the International Monetary Fund or IMF) and at a time of increasing authoritarianism by then President Mugabe. Mugabe's call for the third *Chimurenga* that included forcible land acquisition from the white farmers made Zimbabwe increasingly more confrontational with the West, with Britain in particular, and was accompanied even by name calling, one of which was "The Blair that I know is a toilet" from the song "Agirimende" that became one of the theme songs of the third *Chimurenga* (Nyambi 2017).

By the time Mugabe was removed as president in 2017, the Zimbabwean economy was completely shattered, and the advisor to the new president, Emmerson Mnangagwa, put it up front: "Land reform is over. Now we want inclusiveness" (as quoted in Suzuki 2018: 85). These are perhaps some of the manifestations of the "strange demons from the deep . . . that come trailing all sorts of subterranean material" after the "dismantling of the colonial paradigm," about which Stuart Hall had warned (Hall 1996: 259).

Let me return to Laet and Mol's article. Laet and Mol's erasure of race/racism and the enduring legacies of white-settler colonialism make them slip into the "moral strategy" of the white Zimbabweans.[12]

Although they admit that "[o]ne of the limits of our material is that we do not know how the village women appreciate pushing an iron crossbar with their men sitting on it" (Laet and Mol 2000: 258, note 21), the only actors with names in their story are white men (apart from the Zimbabwe bush pump and briefly and dismissively Mugabe) – namely, Tommy Murgatroyd, Ervin Von Elling, and Peter Morgan. If such erasure does not seem significant, one should simply look at the rallying call of the Black Lives Matter movement in relation to those killed through police brutality – "Say his/her name." These erasures, unfortunately, also make their story seem yet another expression of "European hallucination," the broader goal of which is consolidating "an inside to draw from it some domestic benefit" (Derrida 1998: 87). Laet and Mol present Peter Morgan as the "other" of Louis Pasteur (albeit the Pasteur of Bruno Latour's version) who, unlike Pasteur, instead of consolidating authorship and marshaling the network for his purpose, "seems to dissolve his own actorship" and that, too, "*actively*" (Laet and Mol 2000: 249).

But does Dr Morgan dissolve his actorship or authorship? Or is it that Laet and Mol's bid to show Dr Morgan as the "other" of Louis Pasteur (albeit Latour's version) leads them to present a caricature of his "actorship"? Dr Morgan does not deny his individual role. The website of Aquamor, a "private research and development company based in the city of Harare," whose executive director is Dr Morgan, states clearly: "This website was originally established in 2001 *to describe the work of Peter Morgan who has been involved in research and development within the rural water supply, hygiene and sanitation sector for nearly 40 years*" (emphasis added).[13] In fact, in 1991, six years before Laet and Mol met and interviewed Dr Morgan, he had even received a Member of the British Empire (MBE) award from Queen Elizabeth II. (There are hundreds of people who have declined this award for a variety of reasons, including some for its being a legacy of colonialism).[14] In his book (Morgan 1990), which Laet and Mol "refer to throughout the text as [Morgan]'" (Laet and Mol 2000: 255, note 1), all the photographs are those of Black Zimbabweans, but there is no mention of race/racism or of the impact of colonial history. There is a brief reference to "communal lands" (which emerged from the reserves – the land mandated for the Black Zimbabweans by the settler colonial government). Morgan, citing two studies on water access in the communal lands, observes that "two thirds of the households do not have all year around access to a safe water supply" (Morgan 1990: 67).

I am certainly not trying to discredit Dr Morgan's enormous contributions to public service in Zimbabwe. Rather, I wish to suggest that if Laet and Mol had realistically (and a little more critically) explored Dr Morgan's "actorship" in the making of the Zimbabwe bush pump, perhaps they would have been able to explore the complex and enduring colonial "environment" of the bush pump and its distributed authorship. That may have also allowed Laet and Mol to include the role of the historical and cultural context of Zimbabwe and the "actorship" of Black Zimbabweans beyond women pushing the crossbar and men sitting on it (which unfortunately seems yet another iteration of a common colonial imagery). The water diviners, *ngangas*, are given some "actorship" but their role is presented through a "UNICEF worker," Mr Ervin Von Elling (whose company makes the Vonder rig for the Zimbabwe bush pump), and Dr Morgan (including his book and his letter). The long quote from a UNICEF worker's description of the water diviner's role in Laet and Mol's article makes it unclear whether the power of the diviner is occultic or based on this-worldly expertise as well. In a footnote, drawing on a letter written by Dr

Morgan to them, they highlight the diviner's social and technical expertise: "Drilling wells without consulting a *nganga* would be unwise – not only because when they act as water diviners they know about aquifers, but also because they know about people" (Laet and Mol 2000: 258–9, note 23).

As such, Laet and Mol also lose an opportunity to present a much more inclusive network of human and non-human actors. Mavhunga, for example, shows how the spirituality of *Chimurenga* is "replete with experimentation, application of ideas to practice, and practice generative of new ideas," and is complexly entangled in relationships of humans and animals, inanimate and animate, and linguistic and material (Mavhunga 2017: 58). In fact, Peter Morgan, in the book that is Laet and Mol's main source of information, presents the role of the water diviner very clearly: "The diviner will look at the lie of the land, make a note of the soil type and catchment area, and look for other signs such as '*Muonde*' or '*Mukute*' trees, or lines of anthills to identify a high yielding aquifer" (Morgan 1990: 24). That is, the diviner's expertise presents yet another example of "innovative transformation of *zvakatikomberedza* (surroundings; caves, mountains, rivers, pools, valleys, forests, animals, and trees) into . . . assets and infrastructure, with or without physically modifying them" (Mavhunga 2017: 46). The contrast between this way of being/existence and knowledge making and the colonial approach is stark, as the following description by Mavhunga shows:

During the Second World War, the *British authorities sensationally accused locusts of aiding and abating* [sic] *their enemy*, Adolf Hitler and the Germans. Following the outbreak of hostilities, the British administrators charged that the locusts had "joined the Nazis as enemies of humanity," justifying "campaigns" to be launched against them "from North Africa to India." The biggest "offensive" against this [non-human] Nazi "ally" was in Kenya from 1943 to 1947. It involved "13 drives, the 4th involving 4,000 troops, 33,000 labourers, 750 cars, and 3,000 tons of poison bait in one operation." (Mavhunga 2017: 58; emphasis added)

Sadly, as a result of their erasures, Laet and Mol's idealized construction of Dr Morgan (as an "ideal man") starts to seem like yet another European colonial narrative in which the white man brings salvation to the non-white "others." Such a slippage is not a reflection of individual folly. Rather, it highlights their interpellation through certain ideologies and discourses. And this can, and often is, beyond individual control, particularly if we do not critically engage with our ideological/discursive interpellation. For example, a description of the "actorship" of Dr Morgan in a British media report, after he received the Stockholm Water Prize, gives him a colonial visage – that of the white man saving the world: "*His [Dr Morgan's] work has helped* a large number of the *780 million people worldwide* living without access to safe water *and the 2.5 billion people* who do not have access to adequate sanitation. Diseases linked to such deprivation kill more than 5,000 people each day" (emphasis added).[15]

Portuguese voyages and reframing of the "Great Divide" between the West and the rest

Europe . . . is a contested notion. Historians and critics debate whether it is one of those "inventions" that elites have imposed upon others in order to consolidate positions of power and authority.

J. G. A. Pocock, "What Do We Mean by Europe?," *The Wilson Quarterly*, 1997: 12.

... the creation of the legend of Vasco da Gama [who
commanded the first Portuguese sea mission to India]
began in his own lifetime, and ... he participated in it.

Sanjay Subrahmanyam, *The Career and
Legend of Vasco da Gama*, 1997a: 361

"Columbus's discovery of the New World in 1492,
when taken with the arrival of the heavily armed
Portuguese vessels in the Indian Ocean in 1498, clearly
marks an important turning point in the balance of
power between Europe and the rest of the world" (Law
1986: 234). This opening sentence of John Law's article
"On the Methods of Long-Distance Control" is further
elaborated in the sentence that follows: "From that
moment onwards until the very recent past the rest of
the world has been under European control and domination" (1986: 234). Although Law explains later that
his article's "contribution towards an understanding
of the means of long-distance control involved in the
growth of imperialism" is his "third aim" (the first being
"contribution to a general analysis of long-distance
social control"), as I show in the following, his claim in
relation to the emergence of Europe/the West-versus-the
rest divide overdetermines his argument (Law 1986).[16]
A lack of critical engagement with European colonialism and Orientalism not only limits the scope of Law's
study, it also entraps his analysis within a colonial and
Orientalist framework.

Law poses the question: "How, then, were the
Portuguese able to bombard the Samorin of Calicut, to
fight and win a naval victory against a powerful combined Gujerati and Egyptian fleet at Diu in 1509?" (Law
1986: 236). His answer is "the *Carreira de India*'s" "display [of] certain properties that are crucial to all systems
of long-distance control" (1986: 236). Law suggests:
"To control the Indian Ocean spice trade a structure
with a larger envelope of mobility and integrity was

essential" (1986: 237). "And to do this," Law argues, "it was . . . necessary to control, direct, and maintain the integrity of the vessels sent out from Lisbon" (1986: 237). The Portuguese, according to Law, were able to achieve this because they were able to "discipline" the constitutive elements of long-distance voyages that included the carracks (ships), cannon, navigational techniques, and even the sailors. "The carrack," for example, "was more independent of its environment than earlier types of vessel" (1986: 239). "This capacity . . . [to] extract compliance from the environment," Law argues, "is most vividly exemplified for the case of naval battles or bombardments where unfortunate Muslims or Hindus found themselves at the wrong end of the superior Portuguese cannon-fire" (Law 1986: 240). Law thus concludes, "the Portuguese made good use of documents, devices and drilled people and . . . these were obvious resources to be used by anyone who wanted to exercise long-distance control" (1986: 254).

Law's analysis of the Portuguese sea voyages to India illustrates the broader concern of actor–network theorists. "On the one hand . . . the force of the tables of declination is consistent with and indeed follows Latour's line of reasoning" – in relation to "the importance of the printing press for western social change." "On the other hand, the significance of Latour's analysis of the role of the printing press is that its invention may be seen as a revolutionary improvement in the textual means of long-distance control, *one that goes some way to explaining both the hegemony of the west and the 'great divide' between primitive and modern*" (Law 1986: 255–6; emphasis added). Latour, writing nearly two decades later, acknowledges the importance of Law's study in his book *Reassembling the Social: An Introduction to Actor–Network Theory* (Latour 2005). Latour writes, actor–network theory started in earnest with three documents (Latour 1988; Callon 1986; Law

1986), adding, "[i]t was at this point that non-humans
– microbes, scallops, rocks, and ships – presented themselves to social theory in a new way" (Latour 2005:
10).[17]

The discursive power of the Portuguese "method
of long-distance control" seems to have waned by the
time Law did his study. In Law's description of the
Portuguese voyages, "*Carreira da India* ... the term
used by the Portuguese for the round voyage made
by their Indiamen between Lisbon and Goa from the
time of Vasco da Gama until the substitution of sail by
steam" (Boxer 1960: 35) became *Carreira de India* – the
da of Portuguese gets (mis)translated into the French
de! But let's not get distracted by this relatively trivial
slippage.

Let's return to Law's description of the long-distance
control achieved by the *Carreira da India*. Law writes
that the Portuguese sailors "were ... forceful for the
same two reasons [as was the case with documents
and devices]: first because of the structured envelope in
which they were placed, and second because they themselves embodied a great deal of previous effort" (Law
1986: 253). "The Portuguese mariner, on a vessel with
a cannon," he emphasizes, "was indeed powerful." In
contrast, "[t]he same mariner, *shipwrecked* on a beach,
was pathetically weak" (1986: 253; emphasis added).
Law then goes on to explain the "navigational envelope
... that made mobility and undistorted communication
between Lisbon and India possible" (1986: 251).

Shipwreck is mentioned only once in Law's essay
– in the quote above in which I have highlighted "shipwrecked" – and that too as an exemplification of a
weakness that the Portuguese method of long-distance
control had transcended. Shipwrecks were, however,
quite common. Charles Boxer, a historian, one of whose
works Law cites, for example wrote in 1960: "A long
series of shipwrecks, *arribadas* (abortive voyages), and

other maritime mishaps had reduced the once great and glamorous prestige of the Portuguese Indiamen to a very low ebb by 1650" (Boxer 1960: 35). Boxer recounted how "Richard Flecknoe, who had made a delightful voyage to Brazil and back in 1648–9, was offered a passage to India with the viceroy Count of Aveiras in the next year. After only slight hesitation, he turned this down observing that *'not one Portugal ship of three returns safe from that voyage'*" (Boxer 1960: 35; emphasis added).

An Orientalist imaginative geography that intensifies distance and difference between Europe/the West and the rest (Said 1979) guides Law's analysis and makes him overlook the fact that the definition and consolidation of Europe as an entity itself emerges in the process of these travels, exchanges, and wars. "The 'continent' of Europe," as historian J. G. A. Pocock has shown, "is a product partly of the Mediterranean need for a term to inscribe and describe the lands west of the Bosporus, and partly of the exceptionally self-centered and world-dominating outlook developed by a civilization that evolved in those lands" (Pocock 1997: 14). Pocock shows that the boundaries of Europe remained rather amorphous for a long time; the "process of defining continental Europe was not quite complete" until Voltaire, who celebrated "the world of Peter [the Great] and his successors in bringing Russia into the [European] civilization" (Pocock 1997: 16).

An uncritical presentation of Europe as a separate entity also involves forgetting that Europe is a relatively small part of a massive continental land mass that includes what the Europeans started to call Asia in order to mark Europe's own "distinct and distinguishable identity," which without positing the "other" "is inherently unstable and constantly changing" (Sakai 2001: 791). Such Eurocentrism has often undergirded the narrative of the early Portuguese voyages, which, on the one hand, is evident in the continual efforts – going on until recently – at creating the legend of Vasco da Gama (Subrahmanyam 1997a) and, on the other, in the provincial approach of the Europeans towards global geography. For example, Pierre Chaunu, whose work Law cites as "the most notable exception" to the general lack of historical engagement with "the technology of Portuguese expansion" (Law 1986: 235), continued to "argue that the Indian Ocean was 'scarcely more than an extension of the eastern Mediterranean'" until as late as 1979 (Verne and Verne 2017: 314).

The Orientalist imaginative geography of the West versus the rest also results in erasures of the role of non-Europeans in the early Portuguese voyages. *Carreira da India* depended on the navigational expertise and support of the Africans and the Indians to make the voyage successful. For example, when Vasco da Gama's fleet reached Mozambique on its first voyage to India, "very quickly, relations developed between the Portuguese fleet and the local ruler (Sultan)" and the Portuguese tried to present the Sultan "with hats, coral and the like," which the Sultan declined. Nonetheless, "Vasco da Gama used these contacts to persuade the Mozambique ruler to 'give' him two pilots to take along" because of the realization that they had "entered a world of maritime skills, where it made more sense to use local expertise" (Subrahmanyam 1997a: 112). This was not the only time the Portuguese sought local expertise in navigation. Later, when they reached another port of Africa, Malindi, since, in Mombasa they had lost the pilots who were given to them, the Portuguese asked the local king for a Christian pilot and, as one of the scribes on this fleet noted, "we rejoiced greatly with the Christian pilot whom the King sent us" (1997a: 120).

Another Orientalist trope that runs through Law's description of the early Portuguese voyages to India is in relation to the Muslims (and Islam). While writing about his aim to contribute to an "understanding of means of long-distance control," Law poses a question that he thereafter states he cannot answer: "How was it ... that Christian Europe, at the turn of the fifteenth century, hemmed in in the East by *predatory Muslim powers*, succeeded so dramatically in turning [the] tables?" (Law 1986: 236; emphasis added). Law's argument that "Da Gama's first expedition had carried guns, but more would be needed if the *hostility of the Muslims* was to be mastered" (Law 1987: 246; emphasis added) is intriguing considering he says little about the religious context of these voyages. The trope of the non-European "others" as predators and invaders has been a common colonial trope and, as I show in chapter 2, also undergirds Butterfield's historical account of the origin of the Scientific Revolution in Europe.

Law further reinforces his generalized evaluative references in relation to the Muslims through another question a little later: "How ... were the Portuguese able to ... obtain a stranglehold on the vital Indian Ocean spice trade that had previously been monopolized by Muslim sailors?" (Law 1986: 236). Law, thus, ends up replicating a common Orientalist stereotype that "assailed Islam as if ... [it was] a religio-political culture about which deep generalizations were possible and warranted" (Said 1979: 208). More broadly, Law's study continually reinforces an old Orientalist discursive strategy – "large generalizations by which reality is divided into various collectives" in which "each category [is] not so much a neutral designation as an evaluative interpretation" (Said 1979: 227).

Law's historical-cultural generalizations in relation to trade in the Indian Ocean, at the time when the Europeans entered that trade, start to seem like an Orientalist "imaginative history" (Said 1979) also because his claims are often historically inaccurate. The claim of a Muslim monopoly over the Indian Ocean

trade, for example, presents "the image of a mono-
lith where none existed" (Subrahmanyam 1997a: 96).
"First, other non-Muslim Asian merchant groups participated in this trade on a large scale as well – and often
wholly or largely on their own terms" (1997a: 96).
Moreover, the "Muslims, to exercise a *monopoly*, must
have acted *en bloc*" and that is not demonstrable based
on historical records (1997a: 96).

A critical engagement with Orientalism, Eurocentrism,
and colonialism could have led Law to other possible
interpretations and arguments. For example, to suggest
that one particular event or set of events, such as the
early Portuguese voyages, marked the "turning point"
in the relationship between Europe/the West and the rest
– "*From that moment onwards,*" as Law put it – is itself
a reflection of Eurocentric historicism. There was no
such moment. Well into the sixteenth century, politico-
commercial engagements of the Portuguese remained
contingent on local and trans-local events and agents.
As Sanjay Subrahmanyam shows, "Venice and Flanders
awaited news of the *Carreira da India*, and very often,
this news was less than positive, because ships failed
to complete the voyage ... or because capital for the
pepper and spice cargoes was lacking" (Subrahmanyam
2005a: 36).

More broadly, if we can move beyond the Orientalist
historiography, which made "the history of 'events'"
hostage to "the larger exercise in the history of 'mentali-
ties,'" we can excavate the vibrant connected histories of
such exchanges that cut across Eurocentric dualist boundaries (see, e.g., Subrahmanyam 1997b, 2005a, 2005b). A
critical engagement with the European colonialism and
its multilayered impact would, in fact, allow ANT to
achieve its proposed goal, rather than end up replicating
Orientalist and Euro-/West-centric constructions, as it
often does. To quote Latour, "when the carracks started
to come back with their trophies, booty, gold and spices,

indeed things 'happened' in Lisbon, transforming the little provincial city into the capital of an empire larger than the Roman Empire" (Latour 1987: 230).

Let me return to the opening sentence of Law's article because it too is telling. The phrase "Columbus's discovery of the New World" has definitely become contested, and even jarring, in the present. A recent study has shown that the arrival of Europeans in the Americas resulted in "unprecedented mortality rates" among the Native Americans because of the "introduction of pathogens unknown to the American continent" combined "with warfare and slavery" (Koch, Brierley, Maslin, and Lewis 2019: 20–1). The deaths of the Native Americans from the "Discovery of New World" – the "Great Dying" – was so significant that it *reduced the earth's temperature* by 0.15 degree Celsius (Koch et al. 2019). Not surprisingly, calls to abandon the celebration of Columbus Day have been gaining support in the United States. And eight states (Florida, Hawaii, Alaska, Vermont, South Dakota, New Mexico, Maine, Wisconsin) and also parts of California no longer celebrate Columbus Day.

Law, in starting his article with an imperial claim that hid the violence of coloniality (see Quijano 2000 for a discussion on coloniality) and belied the obvious counterfact – how could anyone discover a place that was already inhabited? – was echoing beliefs that were prevalent at the time his article was published. In 1984, the "Age of Discovery" and Columbus's central role in initiating that age were still largely uncontested in white Euro-America. Interestingly, in the United States, although President Benjamin Harrison "issued a proclamation [in 1892] calling for a new national holiday: 'Discovery Day'" to honor "Christopher Columbus as a symbol of 'progress and enlightenment,'"[18] it was President Franklin Roosevelt, the liberal icon, who formally made Columbus Day a federal holiday in 1937. In

1940, on the eve of the United States joining the Allied forces during World War II, Roosevelt's statement recalled how "[t]he promise which Columbus's discovery gave to the world, of a new beginning in the march of human progress, has been in the process of fulfillment for four centuries."[19]

Nevertheless, a critical engagement with colonialism and Orientalism (and there was already ample literature on these when Law was writing his article, including Edward Said's *Orientalism* [1979]), could have allowed Law to, for example, further explore the genealogies of the concepts of invention and discovery. It could have, for example, allowed him to investigate the colonial genealogy of such usages and to interrogate the Eurocentrism inherent in some of the European claims about technological inventions. As Mario Biagioli (2006) has shown, until the late eighteenth century, when new patent laws were enacted in France and the United States, a patent or privilege in Europe implied a *local novelty*. "It was utterly immaterial whether the inventor had extracted the invention . . . [from] his/her mind or from the country next door" (Biagioli 2006: 1142; Prasad 2014). The fact that this genealogy of European conception of an invention or a discovery is evident only in references to colonization (e.g., the discovery of Americas) should press us to further investigate the colonial genealogies of technological inventions in the late medieval and the early modern eras.

Ironically, a quarter of a century later in 2019, Law (in a co-authored piece), citing his article "On the Methods of Long-Distance Control" (Law 1986), simply writes: "Early ANT *excelled at studies of strategic growth involving European long-distance (and sometimes imperialist) control* . . . in which technologies, natural forces, people, and texts were all (mis)translated for European ends" (Lin and Law 2019: 125; emphasis added). This statement seems ironic because this later article is among his series of interventions in the subfield of post-colonial STS. Yet the colonial dimensions embedded in his own work are simply passed over (in fact, his work is described as having "excelled at studies of strategic growth"), and there is no effort to revise or rethink his own interpellation through Eurocentric and colonial ideology or discourse.

Let me clarify. My interest is not in indicting Law. Rather, as I have argued in this section, a lack of critical engagement with the European colonialism is the reason that ANT, instead of achieving its stated goal of showing the ever-transforming networks in the making of selves, societies, and things, ends up replicating the Orientalist dualist divide of West versus the rest – not just in Law's article that I have discussed, but persistently. Such a lack of critical engagement with the multilayered and enduring impact of European colonialism, however, runs much deeper within the field of STS. In the next section, I show how it results in contradictions in the postcolonial positions of Sandra Harding and Warwick Anderson.

Revisiting postcolonial science studies

> Orientalism is better grasped as a set of constraints upon and limitations of thought than it is simply as a positive doctrine.
>
> Edward Said, *Orientalism*, 1979: 42

Nearly three decades ago, Sandra Harding, in her call to extend the intellectual engagements of science and technology studies (STS), had provocatively posed the question: "Is Science Multicultural?" (Harding 1994, 1998). The multiculturalism that Harding proposed had a new twist, particularly in relation to STS.[20] She argued for an intertwining of STS concerns with postcolonial and feminist ones in order to highlight the multiculturality of modern science.[21] The postcolonial standpoint, for Harding, not unlike the feminist one, aimed at enhancing "the democratic tendencies within the sciences and to inhibit their elitist, authoritarian, and distinctively androcentric, bourgeois, Eurocentric agendas" (Harding 1991: 217).

Warwick Anderson, the other key figure in bringing the "postcolonial" to STS, apart from emphasizing "analytic symmetry and inclusion," has argued that the "postcolonial study of science and technology suggests a means of writing a 'history of the present,' of coming to terms with turbulence and uncertainty of contemporary global flows of knowledge and practice." Later he advocated a shift in focus of postcolonial science studies: "From subjugated knowledge to conjugated subjects" (Anderson 2009). "Conjugated subjects," Anderson wrote in 2009, "is meant to hint at postcolonial hybridity and heterogeneity, suggesting a more complicated and entangled state of affairs" (Anderson 2009: 389). A few years later, Anderson proposed "Asia as method" to conduct postcolonial inquiry (Anderson 2012), which, as he recently elaborated, ties in with his long-standing engagement with STS in East Asia (Anderson 2020b).

Post/decolonial science studies have grown remarkably since the interventions of Harding and Anderson.[22] These studies cannot be reduced to one or two strands of intellectual engagement or to a set of events. Post/decolonial science studies, as several special issues of reputable journals bear witness, has attracted wide-ranging and productive debates (Abraham 2006; Anderson 2002; McNeil 2005; Seth 2009, to name just the first few). Harding has also published a postcolonial science and technology studies reader (Harding 2011b). If we include the workshops, panels for conferences, and, not to forget, individual papers that have focused on post/decolonial science studies or have utilized the analytical category of post/decolonial, we can, albeit with some trepidation, even call it, as Maureen McNeil had presciently suggested, a "movement" that has been engaged in a "reorientation of STS" (McNeil 2005: 105).

In this last section of the chapter, I focus on Harding's and Anderson's interventions, in part because they have been crucial in defining the trajectories of postcolonial science studies but also because their work exemplifies or passes over a contradiction that needs to be addressed much more squarely and critically. This contradiction was highlighted by Itty Abraham way back in 2006. "Is it possible," Abraham had argued, "that postcolonial techno-science can be an alternative mode of analysis at the same time as the postcolonial indexes a locational site for alternative, i.e., non-western knowledges?" (Abraham 2006: 210). Abraham's suggestion was "to see the power of modern science in political terms, as ideology" (Abraham 2006: 210). The issue though is deeper – can we move beyond the Orientalist and colonial tropes and practices if we continue to consider the "other" as an ontological given rather than as historically constituted? That is, the postcolonial indexing of "a locational site for an alternative" is an example of a much broader Orientalist freezing of the "other" that continues to replicate the distance and the difference between the western "self" and the non-western "other," which undergirded the Orientalist imaginative geography (Said 1979).

The indexing of a non-western location for a postcolonial alternative formed the basis for Sandra Harding's intervention. "This geographical metaphor," Harding stated in the introduction to *The Postcolonial Science and Technology Studies Reader* (Harding 2011b), "directs attention to a location, a site in social relations, from which a disadvantaged group learns to observe and speak for itself and to the advantaged group about how unjust and oppressive social relations affect their lives" (Harding 2011b: 19). A postcolonial standpoint, thus, sought to strategically weave together postcolonial location and knowledges to juxtapose "Our story and an Other story" (Harding 1991) and thereby disturb the Eurocentric and positivist constructions of modern science. Harding further elaborated the postcolonial standpoint by suggesting its genealogical link to Marxism, i.e., as "a methodological standpoint [that] arose in Marxian writings about the importance of taking the 'standpoint of the proletariat' to understand how capitalism actually worked" (Harding 2011a: 19).

Harding's position presents a paradox. If we link the postcolonial standpoint to a geographical *metaphor*, as Harding argues, then the role of location in constituting the "standpoint" becomes ambiguous, if not altogether superfluous. And if we were to directly map the postcolonial standpoint to a fixed location(s), albeit strategically, we run the risk of making postcolonial location and identity unitary and frozen in space and time, thereby exoticizing and appropriating not just the non-western "other" but also the western "self."[23] The postcolonial standpoint, even if it is metaphorically mapped onto geography, also makes Harding's own engagement with postcolonialism – as a white person from the West – precarious (and that of many of the authors whose essays are included in the *Reader*). Although she emphasizes that her intervention draws upon affinity and interlinkages between the feminist and the postcolonial standpoints, she does not carefully problematize the notion of affinity and its implications on "standpoint."

Harding seeks to resolve this paradox but ends up further exposing her ambivalence. On the one hand, quoting Fredric Jameson, she suggests that "standpoint theory remains a valuable strategy to articulate the logic of 'a space of a different kind for polemics about the *epistemological priority of the experience of various groups or collectivities*'" (Harding 2011a: 20; emphasis added). On the other hand, soon thereafter she argues, "[s]tandpoint approaches can recognize the positive scientific and political value of local knowledge without falling into claims ... of its legitimacy by only local standards" (2011a: 21). She attempts to undercut the locational indexing of knowledge by stating that the "language I use here of West and non-West is problematic. *It echoes the discredited Orientalism that makes the West the center* of geography, history, and critical analysis" (Harding 2011a; emphasis added). She adds that not only these two terms but also their substitutes, such as "First World–Third World," "developed–under-developed," and so on, are problematic as well. Such dualist framing, she further explains, "homogenizes the two groups and obscures the more complex social relations that exist between and among various global groupings in the past and today" (Harding 2011a: 22, note 1).

The implications of Harding's ambivalence in presenting the role of locally situated experience in producing "standpoints" becomes even more acute when we confront a related concern: Who is going to decide the norms and standards on whose basis legitimacy of knowledge claims can be assessed? Can this not again slip into the colonial practice of white/western people becoming the arbiters of knowledge? These contradictions in Harding's postcolonial standpoint theory cannot be overcome because it treats the "other" as an ontological given. There is no shortcut to critically and rigorously tracing the genealogies of European colonialism and its multilayered and ambivalent impact on identities, histories, political economies, and so on. We would also have to be aware that such interventions can be treacherous, in the sense they can/will reveal (even if not explicitly) our own complicity with colonial discourses. Indeed, even during colonial rule, the colonized were not a singular and homogeneous category in relation to the colonial hierarchy and appropriation. That, however, does not undercut the power and profound impact of European colonialism and its existing legacies.

Anderson does not explicitly tie postcolonial location to the postcolonial analytics that he proposes. He has used a range of analytical and methodological tools to present his postcolonial arguments and to further the field of postcolonial science studies. His desired goal for postcolonial science studies seems to be actor–network analyses that draw upon postcolonial theory to take that extra step: from "actor–network theory almost becomes postcolonial" to actor–network theory becoming postcolonial (Anderson 2009). However, Anderson has not elaborated the implications of his hybrid approach (for example, of combining ANT with Said's Orientalism) and does not explore its inherent contradictions. The issue here is not the inventiveness of ANT in bringing various hybrids to the fold of STS and, more broadly, to that of sociology, anthropology, history, and philosophy. The difficulty arises because ANT does not take into account the history of the present, i.e., the role of history in constituting actors and actions. ANT starts with a *tabula rasa*, and the expressed goal is to then trace the hierarchies by mapping the networks of associations. However, as I have shown through my analyses of two celebrated ANT studies, it continually fails in this regard. And this is not a coincidence. The erasure of the impact of colonialism, racism, gender, and so on in ANT arises *because* of its methodological/ analytical framework.

In the words of Latour, "It is as if we [ANT scholars] were saying to the actors: 'We won't try to discipline you, to make you fit into our categories; we will let you have your own worlds, and only later will we ask you to explain how you came about settling them'" (Latour 2005: 23).

Although ANT's goal is laudable, it needs emphasizing that the actants/actors – the scallops, the door closers, or the inventors and users of the Zimbabwe bush pump – did not choose the ANT practitioners to be their spokespersons. The problem with ANT's approach is that it can easily slip into becoming a "god trick" whereby the spokesperson (the ANT scholar) can end up presenting one's own voice as that of the "other," whose spokesperson s/he claims to be. S/he can also slip into rearticulating the dominant hierarchies and normative values of the time or, worse still, assume the role of a benevolent despot that underlay the European colonial project of civilizing mission. To use Latour's famous example of the door closer, how are we to figure out whether "a door with a powerful spring mechanism" that may make the door "slam in our face" is playing "the role of a very rude, uneducated porter" (Johnson and Latour 1988: 301)? The door closer may be trying to be playful or attempting to lovingly hug, albeit awkwardly. More broadly, as Gayatri Spivak had warned, "[t]he theory of pluralized 'subject-effects' gives an illusion of undermining subjective sovereignty while often providing a cover for this subject of knowledge" (Spivak 1988: 66).

Anderson bypasses a careful and critical engagement with ANT in his presentations of analytics and methodologies for postcolonial science studies. Moreover, in his postcolonial science studies interventions, there is often a slippage into treating the non-western "other" as an ontological given. Recently, in a forum on "Decolonizing Histories in Theory and Practice," which was published in the journal *History and Theory*, Anderson revisits his earlier proposal for "Asia as method" (Anderson 2012). Responding to some of the critiques of his proposal, Anderson writes that his proposal to treat "Asia as method" "seemed . . . a way to decolonize our histories, though it traced a path that *skirted uneasily the dangers of essentialism* and evaded the problems of internal colonialism in the region" (Anderson 2020a: 432; emphasis added). Drawing on Fa-ti Fan's argument, he suggests that "the multiplicity, ambiguity, and elusiveness of 'Asia' is, methodologically speaking, an asset'" (2020a: 432).

Anderson thus skirts engaging with a broader concern – whether his proposal ends up treating the category of Asia as ontologically given, rather than as histori-cally constituted. Fa-ti Fan had in fact responded to Anderson's re-presentation of "Asia as method" by situating its genealogy in "postwar formulation . . . [aimed at] self-criticism of modern Japanese history" (Fan 2016: 361). Fan, highlighting his "reservations about the current scholarship done in the name of 'Asia as method,'" had suggested that "the value of 'Asia as method' derives in part from its eschewing a naive ontology of what 'Asia' or 'East Asia' is, and in part from the possibilities of its critical interventions in discussions of Eurocentrism, imperialism, modernity, and globalization by providing alternative historical accounts of 'East Asia'" (Fan 2016: 363). Therefore, as is evident from the title of his article as well, he gives "*one small cheer* for Asia as method" (Fan 2016; emphasis added).

Anderson's failure to address the pitfalls of indexing knowledge to a particular location in his own interven-tions, barring, for example, passing them over as having "skirted uneasily the dangers of essentialism," unfortu-nately shifts the very nature of the post/decolonial project. I understand that in part his ambivalence in this regard stems from his interest in promoting post/decolonial analytics and methodologies within the field of science and technology studies (STS). Anderson, who is also one of my gurus and a mentor, has often told me that we should let different flowers bloom. I too love flowers. It does make me happy that ANT scholars have started to directly engage with colonialism/postcolonialism. However, I would also like to be alert to Zygmunt Bauman's analysis that shows how post-Enlightenment European concern with "order" finds expression in "gardening" that aimed at removing the "weeds" and as such ended up in, for example, the extermination of the Jews by the Nazis (Bauman 1993).

I am not suggesting that the situation within STS is comparable to what happened in Nazi Germany. My concern is simply that we need to be cognizant of the kind of "garden" that we are creating. Do the inclusions and exclusions of certain flowers without a critical engagement with what kind of a garden is being created make the project of post/decolonial science studies itself toothless or, worse, complicit in colonial strategies? Can a lack of such an engagement also make us slip into colonial and Orientalist entrapments? In fact, as I show in the following, Anderson's own analysis although it seeks not to directly tie postcolonial analytics to "fixed" locations and to move beyond Euro-/West centrism, ends up being, to use a Derridean phrase, "caught in the game."

Let me return to Abraham's claim in relation to post-colonial science studies that I discussed earlier. "When the postcolonial as a mode of analysis is linked to a fixed site of irreducible knowledge claims," Abraham had argued, "it articulates an ontology that ties knowledge to location as a singular and essential quality of place" (Abraham 2006: 210). His concern, as he elaborated soon thereafter, was the failure of postcolonial science studies, and of Anderson in particular, to "see the power of modern science in political terms, as ideology" (2006: 210). Anderson, positioning himself as a postcolonial Latourian, responded: "Perhaps I could have explained better how complexly hybrid, partial and conflicted these proliferating 'modernities' appeared to me" (Anderson 2009: 394). "Abraham's critique," he added, "seems far more relevant to those postcolonial scholars who tend to essentialize ethnosciences or propose certain place-specific thought styles" (2009: 394). Then, in a playful staging of his own postcolonial stance, Anderson turned the tables on Abraham's critique: "Perhaps this is just another example of the need to *provincialize India in postcolonial studies* – or, in Abraham's more general stipulation, the need to situate postcolonial analysis in political and institutional context" (2009: 394; emphasis added).

Albeit written in jest, Anderson's comment is startling. His suggestion that perhaps Abraham's critique reflects "another example of the need to provincialize India in postcolonial studies," not unlike Law, appropriates Chakrabarty's concept of provincializing Europe (though Anderson does not cite Chakrabarty). Does he mean that Abraham in his critique and suggestion to "see the power of modern science in political terms" is presenting an India-centric historicism that consigns the rest of the (postcolonial) world to a "waiting room of history"? This passing reference, without a careful engagement, even though Anderson is just being playful, unfortunately starts sounding like another claim of western and white privilege.

Ironically, Anderson's response to Abraham's critique ends up reinforcing the claim of the latter. In calling for "the need to provincialize India in postcolonial studies," Anderson ends up linking Abraham's critique to the latter's "location." "Writing *from* India," Anderson asserted, "Abraham was acutely aware of the need to decouple questions of ontology from postcolonial science studies, lest we end up cheerleaders for Hindu nationalists asserting their own special modernity" (Anderson 2009: 394; emphasis added). Abraham, when he wrote that essay, was teaching in the United States; in fact, his doctoral education was conducted there as well. An ideal conjugated subject! Nevertheless, Anderson slips into articulating a Eurocentric subjectivity that frames the non-western intellectual as doubly entrenched (in intellectual as well as spatial/cultural/political location), while ignoring his/her own entrenchment.

Interestingly, although Anderson forgot his own "entrenchment," he connected location and identity in relation to another "other." He wrote in the same essay: "Yet Helen Verran, writing with the Yolgnu people in Australia, *might find essentialized indigenous knowledge* performs quite a different politics, working strategically as a *decolonizing rather than a nationalist strategy*" (Anderson 2009: 394; emphasis added). Anderson's statement is not only confusing but also troubling. In his response to Abraham, just before the sentences quoted above, he had argued that the latter's critique "seems far more relevant to those postcolonial scholars who tend to essentialize ethnosciences." And yet soon thereafter he invokes the decolonizing role of "essentialized indigenous knowledge." This is all the more confusing because feminist scholars have already carried out a similar debate critically and productively for a long time (see, e.g., Butler 1990; Haraway 1991) and there have also been extensive postcolonial critiques of the liberal feminists' complicity in the colonial projects of the West (see, e.g., Burton 1994; Nair 1992). Moreover, as Suman Seth pointed out in the same special issue in which Anderson's above-quoted article was also published, "we need to historicize not only the notions and methods of the universal, but those of the particular as well" (Seth 2009: 384).

My intention is certainly not to undermine Anderson's

important contributions to postcolonial science studies and to the history of science and medicine. In a sense, I am positively responding to Anderson's concern that "the failure to activate a critical, not reconciliatory, postcolonialism . . . is not limited to science and technology studies," but there is a "reluctance to recognize and engage directly with the postcolonial spectre haunting globalization" (Anderson 2009: 397). This reluctance, at least in part, could be arising also because of the way the postcolonial project is being framed within the field of science and technology studies (STS). Postcolonial interventions by several prominent STS scholars, as I have shown in this chapter, seem like the claims of a post-racial society, wherein the prefix "post" reflects *transcendence* from colonial/racial binaries. I thus share Anderson's concern with the urgent need to engage with post/decolonial issues. In this chapter, I have explored the genealogies of colonialism within the field of STS in order to initiate a broader and more vigorous dialogue on colonialism, post/decolonialism, and Eurocentrism within STS.

As I complete this book, Afghanistan has fallen back into the hands of the Taliban after twenty years of American intervention (if long-distance control is such an effective strategy, how does one explain the swift fall of Kabul?). In light of the quick fall of Kabul, colonial tropes (various lacks that undergird, as Chakrabarty had argued, Eurocentric historicism, tribalism, lack of a nation, etc.) are in full display in much of the western media to explain why the Afghan army could not resist the Taliban fighters who were largely fighting with AK-47s and some rocket launchers. Ironically, in the present time when the global political economy is radically shifting from West-centric control, the deployment of colonial tropes seems a desperate and futile attempt to hide the coloniality of the American intervention. Colonialism and coloniality are, thus, very much among us, and that too in a wide variety of ways and in a number of forms, as should be evident from the various examples that I have discussed and analyzed in this book.

Conclusion

Modern Science and European Colonialism: A Conversation with J. P. S. Uberoi and Bruno Latour

India as a culture area will be nowhere . . . in the world of knowledge, the sciences and the arts if it does not first defy the European monopoly of the scientific method, established in modern times. It is no solution to propose to wait until we should ourselves become Europeans.

> J. P. S. Uberoi, *The Other Mind of Europe:*
> *Goethe as a Scientist*, 1984: 9

Just as the moderns have been unable to keep from exaggerating the universality of their sciences . . . they have been unable to do anything but exaggerate the size and solidity of their own societies.

> Bruno Latour, *We Have Never*
> *Been Modern*, 1993: 120

J. P. S. Uberoi had started writing about the relationship between science and culture around the same time that science and technology studies (STS) was being born, in the second half of the 1970s.[1] Neither the West nor modern science was alien to Uberoi. He was trained as an engineer at University College London in the 1950s, which resulted in his acquiring "some experience at first hand of the laboratory and the factory in England"

156

(Uberoi 1978: 12). Thereafter, he completed a master's in anthropology at the University of Manchester and then received his PhD in anthropology from the Australian National University. Right from the start of his intellectual career, Uberoi was not an uncritical follower of the western intellectual traditions. His master's work at the University of Manchester had, for example, resulted in one of the first critical reorientations of Bronislaw Malinowski's study of the Kula ring ceremony among the Trobriand Islanders (Uberoi 1971). In relation to modern science, as the first epigraph illustrates, it is important for Uberoi that Indians, and, more broadly, the non-West, defied the "European monopoly of the scientific method" (Uberoi 1984: 9).

Uberoi's concern is not, however, science or scientific method understood in a narrow sense. The intellectual problem that he addresses is much broader: "From our non-Western point of view it is the problem of science and *swaraj*, i.e., of intellectual enlightenment and cultural identity, or, at the individual level, of how to become a modern scientist and still remain oneself" (Uberoi 1978: 12).[2] Although Uberoi chose to speak from "a non-Western point of view," his intellectual and political concerns are not provincial (i.e., they are neither about nor applicable to only non-western societies). He wrote that if his analysis may "seem to be a critique of modern western science and culture, then . . . [he] shall take every care to see that it is a critique offered by someone who is *both an insider and an outsider*" (Uberoi 1978: 12; emphasis added). Uberoi not only resists the Orientalist practice of provincializing (situating the individual and his/her/their thought in his/her/their local contexts) the non-western "other," he exemplifies a cosmopolitanism of the erstwhile colonized which, in contrast to the Orientialist "imaginative geography" that intensifies the difference and distance between the western "self" and the non-western "other"

(Said 1979), does not constitute the West as an ontologically and socio-culturally separate "other."

Uberoi's cosmopolitanism does not imply transcendence of the European colonial discourse of modern science. Uberoi emphasized that "modern scientific and rational knowledge is the self-existent storehouse of truth and it is, *sui generis*, the only one of its kind" and the "rest is charmingly called 'ethnoscience' at best, and false superstition and darkest ignorance at worst" (Uberoi 1978: 14). He does not challenge either modern western science's claim to truth or the characterization of non-western knowledges as ethnoscience.[3] Nonetheless, he shows that modern western science has been historically and culturally situated. Uberoi argues that the origins of modern western science can be genealogically traced to a theological intervention of Ulrich Zwingli, one of the key figures of the Protestant Reformation movement in medieval Europe, because it represented the first articulation of a distinction between fact and value:

> By stating the issue ["whether the true body and blood of Christ are corporeally in the bread and wine"] in terms of a dualism, or more properly a double monism ... [Zwingli] had discovered or invented the modern concept of time in which every event was either spiritual or mental or corporeal and material but no event could be both at the same time. (Uberoi 1978: 28)[4]

This discursive separation of fact and value, according to Uberoi, found a particular expression in the positivist science or, more broadly, in the "positivist regime" that emerged "between the Protestant debate at Marburg, the counter-debate at Trent, and the foundation of the Royal Society of London" and still holds sway in the universities in the West and, through them, in the non-West as well (Uberoi 1978: 53). This genealogy of modern science, which undergirded Newton's linear and dualist theory of colors, represents for Uberoi a "wrong direction" for science. He argued that Goethe's theory of colors – a semiological science that embodied non-dualist knowledge and practice – represented a more desirable genealogy of science for humankind (Uberoi 1984).

Uberoi, although he discursively situates the origins of modern science in history and culture, does not challenge the universality of modern science. Moreover, even though he goes beyond a West-centered genealogy of the "semiological science" (e.g., "Panini, the ancient Sanskrit grammarian" is presented "as the greatest representative of the semiological method" (Uberoi 1978: 17), he does not defy Eurocentric historicism – i.e., "first in Europe, then elsewhere" temporality (Chakrabarty 2000). Uberoi is a structuralist, and the origin of the "structure" of modern science, albeit discursively constituted (and as such not beyond human action), explains the particular practices of science. Nevertheless, despite his structuralist approach, Uberoi, under whom I received my first training in sociology of science, admired Bruno Latour's work and introduced me to it.

After completing my master's at Delhi School of Economics, I started my doctoral study under the guidance of Andrew Pickering, who, in stark contrast to Uberoi, has banished the concept of structure from his vocabulary – the world is a dance of agency for him (Pickering 1995). This does not mean that Pickering is not sensitive to social hierarchies. One of the role models for Pickering, as became evident in our many conversations during my PhD, has been the historian E. P. Thompson, particularly the latter's work *The Making of the English Working Class* (Thompson 1966). In a sense, Pickering, not unlike E. P. Thompson, seems to believe that STS, in particular his writing, "might perform a more poetic redemptive function"

(Satia 2020: 256). Sociological discussions of "structure," for Pickering, failed to acknowledge the agency of those dominated (including the agency of the non-humans) and thereby hid the possibilities of "temporal emergence" and change. However, again in a sense like Thompson, whose "radical approach to history was limited in effect because of his intensely domestic focus on England" (Satia 2020: 256), Pickering's narrow analytical focus (for example, his unidimensional conceptualization of temporal emergence) makes his work limited in understanding the genealogy of colonialism.[5] Interestingly, Pickering, too, admired Latour's work, and we, his students, read almost all the writings of Latour in the courses that Pickering taught. I thus inherited Latourian analytics from two very different intellectual lineages. Before I proceed further, I wish to highlight what might have become evident, that in this concluding chapter I am using my own intellectual genealogy to show how contingently and complexly we are entangled within the discourse of European colonialism.

Bruno Latour, as is evident from the second epigraph, debunks the very concept of modernity and also that of a universal science. Moreover, in contrast to Uberoi's structuralist interpretation, Latour abandons interpretation itself. For Latour, "[e]ither the networks that make possible a state of affairs are fully deployed – and then adding an explanation will be superfluous – or we 'add' an explanation; stating that some other actor or factor should be taken into account, so that it is the *description* that should be *extended* one step further" (Latour 2005: 137). In short, for Latour, human and non-human "actants," to use an actor–network theory (ANT) term, "speak" for themselves through inscriptions, for example, as seen in the role of chlorofluorocarbons in the making of the ozone hole. Our task as analysts – sociologists, historians, anthropologists, geographers, etc. – is simply to *describe* the actions of these agents/actants.

Latour, as the very first sentences of his book *We Have Never Been Modern* (Latour 1993) make clear, described the "inscriptions" of different actants, both human and non-human, that we commonly read about in our newspapers. He, too, like Uberoi, situates himself in relation to an intellectual/political problem, or rather a crisis, that keeps glaring at us, which we have nevertheless chosen to ignore. He describes his reading the newspaper and continually seeing various dualist divides being transgressed in the news reports. He, for example, shows how in a news report "[t]he smallest AIDS virus takes you from sex to the unconscious, then to Africa, tissue cultures, DNA and San Francisco" (Latour 1993: 2). However, Latour observes, "the analysts, thinkers, journalists and decision-makers will slice the delicate network traced by the virus for you into tidy compartments where you will find only science, only economy, only social phenomena, only local news, only sentiment, only sex" (1993: 2).

Latour's concern, at a broad level, is similar to that of Uberoi: "retie the Gordian knot by crisscrossing . . . the divide that separates exact knowledge and the exercise of power – let us say nature and culture" (Latour 1993: 3). However, his approach is radically different from that of Uberoi. Emphasizing the work of Michel Serres and other STS scholars, he suggests relying "on the notion of translation, or network," which is "[m]ore supple than the notion of system, more historical than the notion of structure, more empirical than the notion of complexity" (1993: 3). He argues that the "word 'modern' designates two sets of entirely different practices which must remain distinct if they are to remain effective" – that of translation, which "creates . . . hybrids of nature and culture," and that of purification, which "creates two entirely distinct ontological zones," one of humans and the other of non-humans (1993: 10–11).

The issue, according to Latour, is of directing "our attention simultaneously to the work of purification and the work of hybridization" that "immediately" stops us from "being wholly modern." This shift, however, is not simply about reimagining and relocating our present and with it our future. Rather, we become "retrospectively aware that the two sets of practices have already been at work in the historical period that is ending. *Our past begins to change*" (Latour 1993: 11; emphasis added). In one stroke, Latour erases the history of modernity and reconstitutes our past because:

> if we have never been modern . . . the torturous relations that we have maintained with the other nature-cultures would also be transformed. Relativism, domination, imperialism, false consciousness, syncretism – all the problems that anthropologists summarize under the loose expression of "Great Divide" – would be explained differently, thereby modifying comparative anthropology (Latour 1993: 11–12).

Latour, unlike Uberoi, does not situate himself or his work, except as an individual, i.e., he does not propose to speak from say a western, European, or French point of view. He also does not have to contend with the "double consciousness" that, ninety years before Latour's work, W. E. B. Du Bois (Du Bois 1997 [1903]) had argued marked the state of being of the colonized or the oppressed – as an individual and as the "other" of the western/white "self." Latour speaks as an individual unencumbered by colonial "othering" that, for example, impacted not only Uberoi's intellectual work but also his personhood. Uberoi had confessed: "The relentless logic of this general situation of spiritual travail [as a result of western domination], which has prevailed steadily over the non-Western world . . . inevitably produces in me . . . a shameful inferiority complex" (Uberoi 1978: 14). Uberoi, like any of us, considering the geographical scale and profound impact of European colonialism, is, so to say, a cultural hybrid (which might, and often does, find expression in different forms). Latour does not see/show this hybridity – i.e., his genealogy – as a mixture of European cultural history, culture, class, gender, and so on. Indeed, for him, categories such as class, gender, colonialism, and so on are works of purification, produced by the "modern critical stance."

However, analysts have not only purified the mixture, they have also erased and recrafted it, and those erasures and recrafting bear on our being. This is particularly true of European colonialism (though it can be seen in other forms of domination as well). The oft-quoted claim of the historian J. R. Seeley that the British "seem . . . to have conquered and peopled half the world in a fit of absence of mind" (Seeley 1884: 17) does not simply reflect an empirical or factual error. As Seeley's book *The Expansion of England* (Seeley 1884) shows, Seeley's above-quoted observation is less (if at all) a reflection of the absent-mindedness in the British (and European) colonizing of most of the world than a discursive erasure of the constitutive role of European colonialism that is achieved through historiographic craft, which too is a part of the colonial genealogy.[6] If Latour can operate simply as an individual unencumbered by different elements of "othering" and Uberoi cannot, Latour has to thank historians such as Seeley for that colonial gift.

But how does Latour tell the history of mixtures? For example, if a news report on AIDS weaves together two dramatically disproportionately sized spatial-cultural categories – that of Africa (a continent almost three times larger than Europe) and San Francisco – don't we need to explore the history of this representational mixture? Don't we need to investigate how the insertion of Africa in that mixture presented in a report on AIDS reflects a colonial genealogy? Latour's study of *The Pasteurization of France* (Latour 1988) offers a startling history of hybrids that shows how the insertion of the microbes – the non-human agent – in the historiography of pasteurization allows us a radically different, non-human-centered understanding. What Latour presents is not, however, a new history, but a new ontology. This new ontology is defined through various principles of "irreductions," which Latour describes in Part Two of *The Pasteurization of France* (Latour 1988). These irreductions are aimed at rescuing the agent/actions from the reductions that have been imposed by the analysts. The goal of the new ontology is thus to provide answers to questions such as those that Latour presents in *The Pasteurization of France*:

> What happens when nothing is reduced to anything else? What happens when we suspend our knowledge of what a force is? What happens when we do not know how their way of relating to one another is changing? What happens when we give up this burden, this passion, this indignation, this obsession, this flame, this fury, this dazzling aim, this excess, this insane desire to reduce everything? (Latour 1988: 157)

If we follow these Latourian principles, we can surely present a new ontology of hybrids that continually show transgressions of modern dualist boundaries. However, this description would not simply lack a discussion of power or of the normative order (see, for example, Haraway 1997; Shapin 1998). Rather, history, albeit presented as an artifact of the purifications of the "modernist stance," is itself erased. Consequently, Latour and, more broadly, ANT studies continue to be dependent on the modernist histories and sociologies that they criticize. Let me illustrate my claim with an example. Consider an ANT scholar who comes from Mars – an ideal ANT analyst because s/he/it would (most likely) not know about "sociological" categories such as race, class, gender, etc., and so there is no chance of being a reductionist – and decides to study a slave plantation in the American South. S/he/it can describe the day-to-day functioning of the plantation but cannot provide any explanation for it because that would depend upon an understanding of history (not just of the people but also of the artifacts) and that of social categories such as race/racism. S/he/it may not be able to create an understanding of history simply by putting together temporal slices of descriptions because the Martian may not even have a notion of linear time, which too, after all, is a modernist artifact.

In an ideal ANT scenario, we are left with no basis for explanation because explanation has been banished in favor of description of the "inscriptions" of actants. Ironically, this leaves Latour (and also other ANT scholars, for example, as I showed in chapter 3 in relation to John Law's study of the Portuguese voyages) with no option but to either replicate the existing explanations (e.g., with regard to the Great Divide), even while reconfiguring the agency of the actors (e.g., agency is shown through the practice of making allies and networks rather than from having cognitive difference) or resort to the structuralist strategy of positing an origin (albeit as a non-origin) to provide the explanations of what has happened and is happening.[7] ANT cannot escape historicism (history is itself erased, except as a "critique" of existing historiography and the origin becomes representative of historical time). The nature–culture dualist divide, as the origin myth of modernity, for example, is presented as an explanation, which nevertheless exemplifies Eurocentric historicism: "*Modernization*, although it destroyed the near totality of cultures and natures by force and bloodshed, had a clear objective. Modernizing finally makes it possible to distinguish between the laws of external nature and the conventions of society. The *conquerors undertook this partition everywhere*, consigning hybrids either to the domain of objects or to that of society" (Latour 1993: 130; emphasis added).

In the conclusion to this book, I have engaged with Uberoi's and Latour's radically different approaches to understanding European modernity and how this modernity was born through a belief in a dualist divide to highlight that we cannot bypass the slow and tedious process of mapping the genealogy of European colonialism and its role in the history of the present. There is no holy grail of analytics or methodology that can fully explain different and continued expressions of colonialism. And as the colonial paradigm continues to dismantle, we will have to remain alert to not only the "release [of] strange demons from the deep," about which Stuart Hall (1996) warned us, but also the sediments of the familiar relationships through which our own genealogy has been constituted.

Notes

Preface

1 https://www.thehindu.com/specials/independence-day-india-at-70/our-dolls-house-of-memory/article194910 04.ece, accessed March 2, 2022.
2 Sur argues that "[i]t was not the whims of class ties and old boy networks, as Abraham contends, but rather the fundamental differences in class politics and interest that explain the systematic exclusion of Saha [well-known Indian scientist Meghnad Saha] from all positions of power in the planning and execution of science and technology in India" (Sur 2002: 105).

Introduction: Genealogies of Colonialism in Postcolonial Times

1 https://www.govinfo.gov/content/pkg/CHRG-115 hhrg32689/html/CHRG-115hhrg32689.htm, accessed August 25, 2021.
2 https://techcrunch.com/2021/05/01/is-washington-prepared-for-a-geopolitical-tech-race/, accessed August 25, 2021.
3 https://www.congress.gov/bill/117th-congress/senate-bill/1260/text, accessed August 25, 2021.

4 https://www.whitehouse.gov/briefing-room/state ments-releases/2021/06/08/statement-of-president-joe-biden-on-senate-passage-of-the-u-s-innovation-and-competition-act/, accessed August 25, 2021.

5 https://hbr.org/2010/12/china-vs-the-world-whose-technology-is-it, accessed August 25, 2021.

6 The origin of such articulations is traced to a Frenchman, Renaud Camus, who used the phrase *le grand replacement* and then made it the title of his book that was published in 2012. In the book, Camus argued "Native 'white' Europeans . . . are being reverse-colonized by black and brown immigrants, who are flooding the Continent in what amounts to an extinction-level event." Thereafter, the concern with "replacement" became increasingly prevalent among white supremacist groups all across the world. https://www.newyorker.com/magazine/2017/12/04/ the-french-origins-of-you-will-not-replace-us, accessed August 25, 2021.

7 https://www.govinfo.gov/content/pkg/CHRG-115hhrg32689/html/CHRG-115hhrg32689.htm, accessed August 25, 2021.

8 https://blogs.lse.ac.uk/businessreview/2021/02/26/ how-the-west-can-respond-to-chinas-technology-surge/, accessed August 26, 2021.

9 https://history.state.gov/milestones/1830-1860/china-1, accessed August 27, 2021. "Milestones in the History of US Foreign Relations" was retired in mid-2016, but the "text remains online for reference purposes."

10 https://china.usc.edu/treaty-nanjing-nanking-1842, accessed August 27, 2021.

11 https://www.uscc.gov/sites/default/files/3.10.11Kauf man.pdf, accessed August 27, 2021.

12 I have often referred to Carl Schmitt also because, as emphasized by several scholars, there seems to be a "Schmitt fever" in China that complexly straddles

domestic concerns with Chinese polity and China's international relations, particularly with the West (Libin and Patapan 2020; Toscano 2008).

13 https://www.lse.ac.uk/ideas/publications/reports/ protect-constrain-contest, accessed August 28, 2021.

14 Foucault, following Nietzsche, argues that "the pursuit of origin . . . is an attempt to capture the exact essence of things, their purest possibilities, and their carefully protected identities," which is in stark contrast to the work of a genealogist, for whom there is no essence, and claims to essence are "fabricated in a piecemeal fashion from alien forms" (Foucault 1977a: 142).

15 Pickering uses the concept of temporal emergence to emphasize open-ended becoming that cannot be pre-figured and comes about through the mangle of human and non-human interactions (Pickering 1995). For Foucault, "[t]hat which arises out of emergence cannot exist in-itself – or else the genealogist would be witnessing the birth or origin of a 'thing'" (Wilson 1995: 159).

16 Interestingly, the Royal Society displays a portrait of Robert Clive and lists him within the category of "arts and culture." https://pictures.royalsociety.org/ image-rs-13406, accessed September 1, 2021. The *Encyclopedia Britannica* describes Lord Clive thus: "Clive's talents were outstanding, his character no more unscrupulous than that of many men of his day, and his work marked the real beginning of the British Empire in India." https://www.britannica.com/biogra phy/Robert-Clive/Clives-administrative-achievements, accessed September 1, 2021.

17 For example, "The English East India Company's past has been invoked as evidence before the US Supreme Court on the legality of Guantanamo detainees, as a rhetorical epithet in an international cricket ball-tampering scandal, to contextualize Somali piracy in the Indian Ocean, and . . . as a central villain in the last

two multi-billion dollar Hollywood blockbusters in Disney's *Pirates of the Caribbean* trilogy" (Stern 2009: 1146).

18 Mehta belongs to a family of diamond merchants from Mumbai. The buying of the East India Company was a slow and tedious process for him. He "managed to purchase the company from around 40 stakeholders" and eventually, in 2010, became the major stakeholder after paying US$15 million. https://www.theeastindiacompany.com/press-and-news/the-east-india-company-that-ruled-over-us-for-100-years/, accessed September 2, 2021. I wish to thank Indranil Dutta for bringing to my attention this story of an Indian businessman buying the East India Company.

19 https://www.theguardian.com/world/2017/may/06/east-india-company-british-businessman, accessed September 2, 2021.

20 https://www.theguardian.com/world/2017/may/06/east-india-company-british-businessman, accessed September 2, 2021.

21 https://www.bbc.com/news/world-south-asia-1097 1109, accessed September 2, 2021.

22 https://www.bbc.com/news/world-south-asia-1097 1109, accessed September 2, 2021.

23 https://www.theeastindiacompany.com/about-the-east-india-company-eic/eic-today/, accessed September 2, 2021.

24 https://www.theeastindiacompany.com/about-the-east-india-company-eic/eic-today/, accessed September 2, 2021.

25 https://www.theeastindiacompany.com/about-the-east-india-company-eic/eic-today/, accessed September 2, 2021.

26 https://www.theeastindiacompany.com/about-the-east-india-company-eic/eic-today/, accessed September 2, 2021.

27 https://www.theeastindiacompany.com/about-the-

east-india-company-eic/eic-today/, accessed September 2, 2021.

28 In the last chapter of *The Savage Mind* (1966), Lévi-Strauss, directly engaging with Sartre's conceptualization of analytical and dialectical reason, categorically wrote, "Descartes made it possible to attain universality, but conditionally on remaining psychological and individual; by socializing the Cogito, Sartre merely exchanges one prison for another." He added, "Sartre, who claims to found an anthropology, separates his own society from others" (Lévi-Strauss 1966: 249–50).

29 As such, what was left undisturbed was what Derrida in his analysis of Lévi-Strauss's texts had called the "structurality of structure" that posits a center or "a point of presence" which orients and organizes "the coherence of the system" and guides the "play of the system inside the total form" of the structure (Derrida 1978: 278–9).

30 In the last nearly twenty years, "post/decolonial science studies" has grown into a large subarea within the field of STS. There are a number of studies that explicitly deploy post/decolonial categories in exploring different facets of science and technology (see, e.g., Abraham 2000, 2006; Anderson 2009; Anderson and Adams 2008; Harding 2011b; Irani et al. 2010; Law and Lin 2017; McNeil 2005; Philip, Irani, and Dourish 2012; Redfield 2002; Seth 2009, 2017; Stingl 2016; Verran 2002). There is also a large body of work by scholars in the field of STS that engages with colonialism and science, even if the term post/decolonial may not be directly used (see, e.g., Hecht 2002; Hofmänner 2015; Mavhunga 2014; Philip 2004; Subramaniam 2000). In the short list of citations that I have presented above, I have missed many important works, particularly those from scholars situated in non-western countries, for which I apologize. But I must admit that such an

erasure also highlights my own West-centric location.

Chapter 1 COVID-19, Science versus Anti-Science, and the Colonial Present

1 https://www.npr.org/transcripts/151646558?stor yId=151646558, accessed June 8, 2021. Also see Reilly (2014). Mary Poovey is not suggesting that there are no facts. Instead, she argues that although facts were always contested, there was a relatively high bar for them to appear in print and that has changed with the internet and got worse with the use of mathematical modeling that is used to disseminate information on the internet.

2 As quoted in https://www.who.int/news-room/feature-stories/detail/fighting-misinformation-in-the-time-of-covid-19-one-click-at-a-time, accessed June 18, 2021.

3 https://www.scientificamerican.com/article/the-anti science-movement-is-escalating-going-global-and-kill ing-thousands/, accessed June 7, 2021.

4 https://www.scientificamerican.com/article/the-anti science-movement-is-escalating-going-global-and-kill ing-thousands/, accessed June 7, 2021.

5 file:///Users/amit/Desktop/journal.pbio.3001068.pdf, accessed June 7, 2021.

6 https://www.cnn.com/2020/06/28/health/fauci-coronavirus-vaccine-contact-tracing-aspen/index.html, accessed June 7, 2021.

7 https://www.nature.com/articles/d41586-021-01031-w, accessed June 7, 2021.

8 https://www.deccanchronicle.com/opinion/columnists/210521/patralekha-chatterjee-the-rise-of-anti-science-as-covid-19-cases-exp.html, accessed June 7, 2021. https://www.bbc.com/news/blogs-trending-56845610, accessed June 7, 2021.

9 https://www.nature.com/articles/d41586-021-01166-w,

accessed June 7, 2021.

10 World Health Organization. Novel coronavirus (2019-nCoV) Situation Report – 13. https://www. who.int/docs/default-source/coronaviruse/situation-reports/20200202-sitrep-13-ncov-v3.pdf?sfvrsn=195 f4010_6, accessed June 11, 2021.

11 https://languages.oup.com/word-of-the-year/2016/, accessed June 8, 2021.

12 https://languages.oup.com/word-of-the-year/2016/, accessed June 8, 2021.

13 Timothy Caulfield, who has "studied the spread and impact of health misinformation for decades," noted in an article in *Nature* that he has "never seen the topic taken as seriously as it is right now." https://www. nature.com/articles/d41586-020-01266-z, accessed June 11, 2021.

14 The "science wars," which started with the publication of Paul Gross and Norman Levitt's book (Gross and Levitt 1994), resulted in a vigorous debate within and outside STS about the nature of science and its broader impact on society and the role of social scientists and humanities scholars (Fujimura and Luce 1998; Jasanoff 2000; Kleinman 1998). The critique of STS during science wars was also given a postcolonial twist (see, e.g., Nanda 1997).

15 https://www.theguardian.com/world/2019/jun/20/ donna-haraway-interview-cyborg-manifesto-post-truth, accessed June 11, 2021. Similarly, Bruno Latour, when asked about the "science wars," replied, "I certainly was not anti-science, although I must admit it felt good to put scientists down a little." Latour reiterated in the interview that science needs to be defended, but as "science in action" (Vrieze 2017: 159).

16 David Bloor presented "causality, impartiality, symmetry, and reflexivity" as the "four tenets" of the "strong programme in the sociology of knowledge," or the sociology of scientific knowledge (SSK), which later

became the guiding principles of STS. Bloor argued for a principle of symmetry in the analysis of true and false beliefs (Bloor 1991). Actor–network theory (ANT) further extended the principle by emphasizing symmetrical treatment of human and non-human agency (Callon 1986; Callon and Law 1995; Johnson and Latour 1988; Latour 1987; Law 1992).

17 Warwick Anderson has lamented the lack of traction of postcolonial approaches within STS. However, he added that "[w]hile they [actor–network theorists] shy from any direct engagement with postcolonial studies, they seem to have picked up and amplified the vibe" (Anderson 2009: 392).

18 https://factcheck.afp.com/false-reasons-refuse-covid-19-vaccines-circulate-online, accessed June 14, 2021.

19 https://factcheck.afp.com/false-reasons-refuse-covid-19-vaccines-circulate-online, accessed June 14, 2021.

20 Historians have analyzed rumors and misinformation and their role as sources for historical analysis for a long time. More recently, subaltern historians documented how rumor as "the trigger and mobilizer" became an "instrument of [peasants'] rebel transmission" (Guha 1983: 256) and for creating the aura of nationalist leaders such as Gandhi (Amin 1995). Rumors have also been studied to excavate the suppressed tropes and voices of the erstwhile colonized (Stoler 2010a; White 2000a, 2000b). For a review of the literature on "the role of rumour in history writing," see Ghosh (2008).

21 https://www.bitchute.com/video/TuzSyJkjvS4d/, accessed September 15, 2020. Transcription of the interview mine.

22 https://www.nytimes.com/2020/05/20/technology/plandemic-movie-youtube-facebook-coronavirus.html, accessed June 14, 2021.

23 https://patrio.tv/watch/plandemic_F9yIyW5mdRPtax

U.html, accessed June 14, 2021.

24 https://www.amazon.com/Plague-Corruption-Restoring-Promise-Childrens/dp/1510766588/ accessed June 14, 2021.

25 https://www.bbc.com/news/world-australia-411046 29, accessed June 14, 2021.

26 https://www.amazon.com/Kent-Heckenlively/e/B00 J08DNE8/ref=dp_byline_cont_pop_book_2, accessed June 14, 2021.

27 https://www.youtube.com/watch?v=ZNT-aJNtg44, accessed June 17, 2021. Transcription of the sermon in the video is mine.

28 https://www.amazon.com/Plague-Corruption-Restoring-Promise-Childrens/dp/1510766588/ #customerReviews, accessed June 17, 2021.

29 https://www.youtube.com/watch?v=ZNT-aJNtg44, accessed June 17, 2021.

30 https://www.patheos.com/blogs/hedgerow/2020/05/ rev-danny-jones-gives-one-of-the-best-explanations-about-the-coronavirus-shutdown-in-roughly-30-min-videos/, accessed June 17, 2021.

31 https://www.northsidesun.com/columns/one-world-conspiracy-200505, accessed June 17, 2021.

32 https://news.harvard.edu/gazette/story/2020/10/what-caused-the-u-s-anti-science-trend/, accessed June 14, 2021. For Skocpol's study on opposition to climate science in the United States, see Skocpol 2014.

33 https://theconversation.com/covid-19-anti-vaxxers-use-the-same-arguments-from-135-years-ago-145592, accessed July 20, 2021.

34 https://www.pewresearch.org/science/2019/08/02/ trust-and-mistrust-in-americans-views-of-scientific-experts/, accessed June 18, 2021.

35 https://www.pewtrusts.org/en/trend/archive/winter-2021/why-we-must-rebuild-trust-in-science, accessed June 18, 2021.

36 https://fivethirtyeight.com/features/most-americans-

havent-stopped-trusting-scientists/, accessed June 18, 2021.

37 https://fox17.com/news/local/tennessee-senator-calls-out-big-tech-in-fauci-feud-says-shes-standing-up-for-science-marsha-blackburn-dr-anthony-fauci-facebook-twitter-youtube-coronavirus-covid19-pandemic; https://www.thewrap.com/tucker-carlson-says-dr-fauci-should-be-investigated-cites-fringe-lies-about-covid-19-origins-video/; https://www.latimes.com/business/story/2021-06-08/fauci-attack-slogans, accessed June 18, 2021.

38 https://www.cnbc.com/2021/06/09/fauci-blasts-preposterous-covid-conspiracies-accuses-critics-of-attacks-on-science.html, accessed June 18, 2021.

39 https://fox17.com/news/local/tennessee-senator-calls-out-big-tech-in-fauci-feud-says-shes-standing-up-for-science-marsha-blackburn-dr-anthony-fauci-facebook-twitter-youtube-coronavirus-covid19-pandemic, accessed June 18, 2021.

40 "QAnon is the umbrella term for a set of internet conspiracy theories that allege, falsely, that the world is run by a cabal of Satan-worshiping pedophiles." QAnon started in October, 2017 with a post that "appeared on 4chan . . . from an anonymous account calling itself 'QClearance Patriot.'" QAnon has brought together a range of conspiracies, and even some US Congress members have expressed support for it. https://www.nytimes.com/article/what-is-qanon.html, accessed June 17, 2021.

41 https://www.isdglobal.org/isd-publications/qanon-and-conspiracy-beliefs/#, accessed June 17, 2021.

42 Ian Hacking, drawing on Foucault's concern with the history of the present and, more broadly, with the latter's explorations of the entanglements of knowledge, power, and ethics, presents his concept of historical ontology that emphasizes "the way in which the possibilities for choice, and for being, arise in history"

(Hacking 2004: 23).

43 https://m.facebook.com/cthagod/photos/a.16884983
24529881/3356644221048608/?type=3&source=57,
accessed June 19, 2021.

44 *The Breakfast Club* was started in 2010 and it is
co-hosted by Charlamagne Tha God, DJ Envy, and
Angela Yee. The show has "8 million monthly lis-
teners and ... more than 3.5 million subscribers
on YouTube." It has been touted as a "must-listen
radio for racial reckoning." https://www.nbcnews.
com/politics/2020-election/breakfast-club-radio-show-
emerges-crucial-stop-2020-democrats-n991576;
https://www.latimes.com/entertainment-arts/business/
story/2020-07-15/the-breakfast-club-radio-racial-reck
oning-charlamagne, accessed June 20, 2021.

45 https://mobile.twitter.com/cthagod/status/129312940
3170525184?lang=en, accessed June 20, 2021.

46 https://revealnews.org/article/where-did-the-micro
chip-vaccine-conspiracy-theory-come-from-anyway/,
accessed June 19, 2021.

47 https://www.theguardian.com/world/2020/oct/26/
survey-uncovers-widespread-belief-dangerous-covid-
conspiracy-theories, accessed June 20, 2021.

48 https://www.npr.org/2020/12/24/950102024/mis
information-spread-by-anti-science-groups-endangers-
covid-19-vaccination-effo, accessed June 21, 2021.

49 https://www.nature.com/articles/d41586-021-
01084-x, accessed June 21, 2021.

50 https://www.usatoday.com/story/opinion/todaysdebate/
2021/06/02/covid-counteract-anti-vaccine-anti-science-
aggression/5267804001/, accessed June 21, 2021.

51 In the United States, for example, towards the end
of June, 2021, when President Joe Biden's administration was working hard to reach the goal of fully
vaccinating the American people, Associated Press
found that "almost all recent deaths from coronavirus in the US" were "among those who ... [had]

not been vaccinated." https://thehill.com/policy/
healthcare/560240-almost-all-us-coronavirus-deaths-
among-unvaccinated, accessed June 27, 2021.
52 https://www.frontiersin.org/articles/10.3389/fpsyg.
2021.646394/full, accessed June 20, 2021.
53 https://www.nature.com/articles/s41599-021-007
81-2, accessed June 20, 2021.
54 Ajay Sethi, a professor at the University of Wisconsin-
Madison's School of Medicine and Public Health
and director of its Master of Public Health Program,
who has taught a course titled "Conspiracies in
Public Health" for several years, for example, suggested, "Fear is a source of conspiracism. Uncertainty.
Not feeling control." https://www.wiscontext.org/
anxiety-hope-trust-and-slowing-spread-covid-19-con
spiracy-theories, accessed July 19, 2021.
55 https://www.france24.com/en/20200301-with-only-
three-official-cases-africa-s-low-coronavirus-rate-puz
zles-health-experts, accessed July 16, 2021.
56 https://graphics.reuters.com/world-coronavirus-track
er-and-maps/regions/africa/, accessed July 16, 2021.
57 http://www.cameroonconcordnews.com/did-chinese-
doctors-confirm-african-people-are-genetically-resis
tant-to-coronavirus/, accessed July 16, 2021.
58 https://www.reuters.com/article/uk-factcheck-
coronavirus-ethnicity/false-claim-african-skin-resists-the-
coronavirus-idUSKBN20X27G, accessed July 16, 2021.
59 https://www.bloomberg.com/news/articles/2020-03-
14/no-black-people-aren-t-immune-to-covid-19,
accessed July 16, 2021.
60 https://www.cdc.gov/coronavirus/2019-ncov/covid-
data/investigations-discovery/hospitalization-death-
by-race-ethnicity.html, accessed July 17, 2021.
61 https://www.bloomberg.com/news/articles/2020-
03-14/no-black-people-aren-t-immune-to-covid-19,
accessed July 16, 2021.
62 https://www.kqed.org/news/11861810/no-the-tusk

egee-study-is-not-the-top-reason-some-black-ameri
cans-question-the-covid-19-vaccine, accessed July 16,
2021.

63 https://www.cdc.gov/coronavirus/2019-ncov/covid-
data/investigations-discovery/hospitalization-death-
by-race-ethnicity.html, accessed July 17, 2021.

64 https://texaspolitics.utexas.edu/polling/search/year/
2020/month/10/topic/coronavirus-716?page=3#race;
https://texaspolitics.utexas.edu/polling/search/year/
2021/month/02/topic/coronavirus-716?page=2#race,
accessed July 17, 2021.

65 https://www.texastribune.org/2021/03/23/covid-
vaccine-hesitancy-white-republicans/, accessed July 17,
2021.

66 Fredric Jameson gives Kevin Lynch's (1960) concept
of cognitive mapping a Marxist slant by arguing that
an individual's cognitive map reflects his/her interpellation within the capitalist ideology and, through that,
his/her position within late capitalism (Jameson 1988,
1992).

67 https://m.facebook.com/cthagod/photos/a.16884983
24529881/3356644221048608/?type=3&source=57,
accessed July 17, 2021 (publicly available).

68 https://www.youtube.com/watch?v=qd70IwQRhxQ,
accessed July 20, 2021.

69 https://www.nbcnews.com/politics/2020-election/
breakfast-club-radio-show-emerges-crucial-stop-2020-
democrats-n991576, accessed June 25, 2021.

70 https://www.theverge.com/22516823/covid-vaccine-
microchip-conspiracy-theory-explained-reddit,
accessed July 22, 2021.

71 https://www.theverge.com/22516823/covid-vaccine-
microchip-conspiracy-theory-explained-reddit,
accessed July 22, 2021.

72 https://www.theverge.com/22516823/covid-vaccine-
microchip-conspiracy-theory-explained-reddit,
accessed July 22, 2021. See also https://static1.square

space.com/static/5f7671d12c27e40b67ce4400/t/6
0a3d7b3301db14adb211911/1621350327260/
FINAL+for+posting_Facebook+Survey+Summary+
Document+for+Website.docx.pdf, accessed July 22,
2021.
73 https://teachingamericanhistory.org/library/document/
national-life-and-character/, accessed July 27, 2021.
74 https://www.theverge.com/22516823/covid-vaccine-
microchip-conspiracy-theory-explained-reddit,
accessed July 22, 2021.
75 David Arnold has shown how, in the context of
colonial India, western medicine, albeit contested,
undergirded a British colonial biopolitical strategy that
"used – or attempted to use – the body as a site for the
construction of its own authority, legitimacy, and control" (Arnold 1993: 8).
76 Bourdieu argues that when the doxa, which operates
outside the discourse because of its naturalization, is
challenged through the emergence of heterodoxy that
opposes the dominant discourse, it results in a crisis
as a result of the dominant groups countering heterodoxy through orthodoxy (Bourdieu 1977). For the
conceptualization of the nation as an "imagined community," see B. Anderson (1991). Eugen Weber shows
that before the twentieth century there was no singular Frenchness through which the French nation was
imagined as a community. The French nation was born
through the imposition and expansion of the dominant
Parisian culture, and this process, in significant ways,
was akin to colonialism (Weber 1976).
77 https://www.openaccessgovernment.org/muslim-
population-covid/110345/, accessed July 18, 2021.
78 https://www.survivalinternational.org/news/12535,
accessed July 18, 2021.
79 https://www.bbc.com/news/blogs-trending-56919424,
accessed July 18, 2021.
80 https://theconversation.com/donald-trumps-chinese-

virus-the-politics-of-naming-136796, accessed July 28, 2021.

81 https://www.nytimes.com/2020/03/18/us/politics/ china-virus.html, accessed July 28, 2021. In the United States, in the second half of the nineteenth century the "yellow peril" became "a popular term . . . to warn that Japanese and Chinese hordes were on the way to take over white Americans and destroy the white civilization" and, along with negative depictions of the East Asians in the media, comics, and, more broadly, in the public discourse, racial violence was normalized and the exclusion of Asians became a part of government policies (Shim 1998: 387–8).

82 https://www.whitehouse.gov/briefing-room/statements-releases/2021/05/26/statement-by-president-joe-biden-on-the-investigation-into-the-origins-of-covid-19/, accessed July 27, 2021.

83 https://www.wsj.com/articles/intelligence-on-sick-staff-at-wuhan-lab-fuels-debate-on-covid-19-origin-11621796228, accessed July 30, 2021. See also https://www.wsj.com/articles/intelligence-on-sick-staff-at-wuhan-lab-fuels-debate-on-covid-19-origin-116 21796228, accessed July 30, 2021.

84 https://www.nature.com/articles/d41586-021-013 83-3, accessed July 30, 2021.

85 https://www.vanityfair.com/news/2021/06/ the-lab-leak-theory-inside-the-fight-to-uncover-covid-19s-origins, accessed July 30, 2021.

86 https://www.smh.com.au/national/eat-a-bat-and-die-vile-threats-against-wuhan-lab-conspiracy-buster-20210701-p5861i.html, accessed July 30, 2021.

87 https://www.reuters.com/investigates/special-report/ health-coronavirus-who-tedros/, accessed July 29, 2021.

88 https://www.who.int/director-general/speeches/detail/ who-director-general-s-remarks-at-the-member-state-briefing-on-the-report-of-the-international-team-

studying-the-origins-of-sars-cov-2, accessed July 29, 2021.

89 https://www.nature.com/articles/d41586-021-015 29-3, accessed July 30, 2021.

90 https://www.nature.com/articles/d41586-021-013 83-3, accessed July 30, 2021.

91 https://www.nature.com/articles/d41586-021-015 29-3, accessed July 30, 2021.

92 https://www.cnn.com/2021/06/13/politics/g7-communique-china/index.html, accessed July 29, 2021.

93 https://www.worldometers.info/coronavirus/, accessed July 29, 2021.

94 https://foreignpolicy.com/2021/06/11/vaccine-diplomacy-boosts-china-in-latin-america/, accessed July 30, 2021.

95 https://foreignpolicy.com/2021/07/09/biden-china-vaccine-diplomacy-coronavirus-latin-america-covid-19/, accessed July 30, 2021.

96 https://critinq.wordpress.com/2020/03/26/is-this-a-dress-rehearsal/, accessed July 29, 2021.

97 https://www.nobelprize.org/prizes/medicine/1958/summary/, accessed November 13, 2020.

98 https://asiatimes.com/2021/08/chinas-zero-covid-policy-could-be-self-defeating/, accessed August 14, 2021.

99 https://www.connexionfrance.com/French-news/Disputed-French-Nobel-winner-Luc-Montagnier-says-Covid-19-was-made-in-a-lab-laboratory, accessed March 29, 2021.

Chapter 2 Historicism without History: The Scientific Revolution, Reimagining the European Past, and Postcolonial Futures

1 https://nacla.org/article/latin-america-science-long-view, accessed September 24, 2021.

2 https://ncert.nic.in/desm/pdf/phy_sci_partI.pdf, accessed September 24, 2021.

3 I. B. Cohen argues that the phrase "scientific revolution" has been used in various senses at least since the eighteenth century (Cohen 1976). Butterfield was not who "introduced the expression 'the Scientific Revolution' into historical discourse." However, he admits that "Butterfield stressed the revolutionary consequences of what he called putting on 'a different kind of thinking-cap'" and "was largely responsible for making the Scientific Revolution a central issue in the mind of every reader" (Cohen 1985: 390). Similarly, H. Floris Cohen has argued that "Butterfield's usage of the term 'Scientific Revolution' as a central category in his historical survey gave wide 'currency to the name' for the first time, even though he had not been the first to use it" (H. F. Cohen 1994: 112).

4 There were also disciplinary tensions between Needham and Butterfield and their followers with regard to the goal and focus of the history of science department at the University of Cambridge. "In 1936 Needham and Pagel had set up a History of Science Committee, but their departure from Cambridge ... during World War II allowed control of the committee to pass into the hands of a coterie of humanities scholars led by Herbert Butterfield." This change "made 'a rather depressing impression' on Needham when he returned in 1948" and "theologian Charles Raven wrote ... to Needham" that there "was the 'real danger that the History of Science may become a convenient refuge for second-rate scientists'" (Falk 2014: 113).

5 Joseph Larmor (1857–1942), an Irish physicist, was the first to calculate the rate at which energy is radiated by an accelerated electron, and the first to explain the splitting of spectrum lines by a magnetic field. He was knighted in 1909 and "represented his university in the British Parliament from 1911 to 1922." https://www.

britannica.com/biography/Joseph-Larmor, accessed February 4, 2022.

6 According to Jan Golinski, "the modern conception of science as a singular and potentially unified entity was forged" in "the early nineteenth century" (Golinski 2012: 21).

7 For Sarton, "Auguste Comte must be considered as the founder of the history of science, or at least as the first who had a clear and precise, if not a complete, apprehension of it" (Sarton 1916: 330).

8 Well-established western (and also non-western) scholars such as Bertrand Russell, Filmer Northrop, or Wilmon Sheldon, to name a few, commonly used science to demarcate the difference between "the West and the Rest" well into the twentieth century (see Hart 1999 for some such examples).

9 Basalla's article spawned a productive debate that continues to this day. In a recent article, Warwick Anderson historically situates the emergence of Basalla's diffusion model and the critical engagements with it and argues that "changing epistemological premises, especially inclinations towards heterogeneous local framings of science, caused his model to become incongruous and defunct. In the new intellectual environment, his evolutionary theory of the spread of western science had to be decommissioned" (Anderson 2018: 78).

10 According to Priya Satia: "The idea of history as 'something that equally comprises past and future states of a continuous subject, so that we may speak of the history of history as such,' emerged in the second half of the eighteenth century" (Satia 2020: 16). In significant ways "modern science" provided an ideal foundation for this western imaginary of the role of history.

11 In this regard, Butterfield emphasizes the role of the "men of letters" as even more significant than that of the scientists. The Republic of Letters played a very important role in Europe. According to Daston, the

Republic of Letters stood for virtues that we commonly ascribe to modern scientific practice. It espoused cosmopolitanism and impartiality that was to be achieved through detachment – "detachment from religious fervor, from aristocratic favors, even from family and friends – all in the name of impartiality and disinterestedness" (Daston 1991: 382).

12 https://www.routledge.com/The-Origins-of-History/ Butterfield-Watson/p/book/9781138187832, accessed September 28, 2021.

13 Butterfield argued: "In both fields the Chinese themselves have had to become the pupils of the West." He also argued that "[s]ome civilisations, like that of India, remained curiously unhistorical" (Butterfield 1981: 13).

14 https://www.nytimes.com/2008/06/08/books/review/ Becker-t.html, accessed September 13, 2021. The reviewer is particularly concerned with these repetitions or factual errors because Needham was "renowned for his photographic memory, countenancing such slips."

15 Amitav Ghosh, a novelist who was trained as an anthropologist, transposes two historical times, one in the late 1970s, when he was conducting his fieldwork in Egypt, when Egyptians described India/Indians through the Orientalist discourses (that is the case with the Indians too) and the other in the twelfth century, when a Jewish/Arab businessman traveled to Malabar, India, got married, had children and lived there for nearly two decades before returning to Egypt (Ghosh 1992). The connected histories of Egypt/Africa and India, such as those described by Ghosh, do not exist in the popular imaginations of either the Indians or the Egyptians; instead, what we have is information and imaginaries filtered through the European writings.

16 In relation to historians writing about colonialism, Priya Satia has argued that a "focus on intentions presumes active, unmediated conscience" (Satia 2020: 6).

17 In this quote, Raina is specifically referring to an article that Needham had "written under the pseudonym Holorenshaw ... about himself" In this article, "Holorenshaw referred to Needham [that is himself] as an 'Honorary Taoist'" (Raina 1995: 1906).

18 David Arnold presents the idea of "Nehruvian science as a way of framing the problem of postcolonial science in relation to ... India's ... scientific standing in the modern world, and the extent of its scientific ambitions under colonialism and after" (Arnold 2013: 360–1).

19 According to Peter Dear, "the natural-philosophical dimension ["within Needham's project"] is treated as largely epiphenomenal but is not entirely ignored – presumably because it resembles the natural-philosophical aspects of Western science" (Dear 2005: 405).

20 His Eurocentrism is also evident in his relegation of Asia/the non-West to the "Asiatic mode of production" in explaining why modern science did not develop outside Europe (Raj 2016).

Chapter 3 Colonialism and Euro-/West Centrism: Postcolonial Desires, Colonial Entrapments

1 Law and Lin suggest that there are some exceptions and they cite Warwick Anderson's (Anderson 2008), Judith Farquhar's (Farquhar 2002; Farquhar and Zhang 2012), and Mei Zhan's (Zhan 2009) studies as the exceptions.

2 In Mbembe's imaginary of "one world," engagement with the other is presented thus: "This presence for the Other, alongside the Other, as a witness for the Other, is another name for the politics of the gift, of obligation, of freedom" (Mbembe 2017: 176).

3 Chakrabarty's "provincializing Europe" has been

deployed by several STS scholars (see, e.g., Anderson 2009; Prasad 2014; Redfield 2002; Seth 2009).

4 See Mbembe's talk titled "Negative Messianism in the Age of Animism," at the Institute of the Humanities and Global Cultures. https://www.youtube.com/watch?v=kyHUJYfk_os, accessed August 9, 2021.

5 They, however, take a step forward from earlier European/Orientalist constructions of the non-western intellectual. As Veena Das (1984) has shown, Louis Dumont's criticism of anthropologist A. K. Saran presented the latter as "doubly entrenched" – as an Indian/Hindu and as an anthropologist, while forgetting his own entrenchment. Law and Lin are symmetrical in presenting such double entrenchments.

6 Derrida, while analyzing Leibniz's fascination with the philosophical elements of the Chinese script, described such fascination as an ethnocentrism that reflects "European hallucination," Derrida argues, "far from proceeding, as it would seem, from ethnocentric scorn, [such a European fascination or hallucination] takes the form of hyperbolical admiration" (Derrida 1998). For a critique of Derrida's own entrapment within European ethnocentrism, see, for example, Gayatri Spivak's preface to *Of Grammatology* (1998), Rey Chow (Chow 1993), and Sean Meighoo (Meighoo 2008).

7 According to Shapin, "the 'dog' that – so to speak – 'doesn't bark' in Latour's picture of scientific travel is a conception of normative order" (Shapin 1998: 7). He argues for a focus on the role of "trust" in scientific practice.

8 https://www.romfordrecorder.co.uk/news/from-gidea-park-to-africa-peter-s-fantastic-journey-2995804, accessed August 11, 2021.

9 https://www.thezimbabwean.co/2011/12/blair-research-laboratory-renamed/, accessed August 11, 2021.

10 https://www.aljazeera.com/opinions/2020/9/20/who-

is-to-blame-for-zimbabwes-land-reform-disaster, accessed August 11, 2021.

11 For a brief biography of Dr Peter Morgan, see https:// www.siwi.org/prizes/stockholmwaterprize/laureates/ 2013-2/2013-laureate-bio/, accessed August 12, 2021.

12 Mahvunga, while criticizing post-independence Zimbabwean politicians, without singling out Mugabe, puts it bluntly: "From the perspective of the Rhodesians (white settlers of Rhodesia and their descendants), both as a legitimate anticolonial struggle and a historical genealogy *chimurenga* did not exist. The Rhodesians can't even accept that the black person has a brain, and our politicians' departure from the ethics of communal knowledge, purpose, and action that undergirded *chimurenga* have helped 'prove' their point" (Mahvunga 2017: 47).

13 http://aquamor.info/index.html, accessed August 12, 2021.

14 For example, in 2019 George the Poet (George Mpanga), in his rejection of an MBE, stated: "I see myself as student, admirer and friend of Britain. However, the colonial trauma inflicted on the children of Africa, entrenched across our geopolitical and macroeconomic realities, prevents me from accepting the title Member of the British Empire." https://www. theguardian.com/politics/2019/nov/25/george-the-poet-rejected-mbe-pure-evil-british-empire, accessed August 12, 2021.

15 https://www.romfordrecorder.co.uk/news/from-gidea-park-to-africa-peter-s-fantastic-journey-2995804, accessed August 12, 2021.

16 In another article, published a year later, Law describes in greater detail how the "disciplining" of navigation was achieved by the Europeans, the Portuguese in particular, in the fifteenth century (Law 1987).

17 The three "documents" that Latour is referring to are his study of "Pasteurization of France" (Latour 1988),

Michel Callon's study of scallops (Callon 1986), and John Law's "On the Methods of Long-Distance Control" (Law 1986).

18 https://www.holycross.edu/sites/default/files/educ/columbus_chapter.pdf, accessed August 15, 2021.

19 https://www.presidency.ucsb.edu/documents/statement-columbus-day-0, accessed August 15, 2021.

20 Interestingly, Harding's argument not only initiated a debate within STS, it also became a target in the "science wars." Meera Nanda, for example, argued: "For Harding, as for many other advocates of multicultural science, the impulse to empathize with non-western 'Others' requires that knowledge systems not be rank-ordered in terms of better or worse accounts of reality" (Nanda 1997: 81).

21 "Postcolonial analyses," she had concluded, "are exactly what is called for by its conventional goal of increasing the growth of knowledge through critical examination of cultural superstitions and unwarranted assumptions" (Harding 1994: 330).

22 There has also been a productive debate on whether "postcolonial" or "decolonial" is a more appropriate term reflecting the diverse engagements with the impact of colonialism across the world (see, e.g., Anderson 2020a).

23 It also runs the risk of "mapping . . . difference onto an underlying hegemony" (L. Cohen 1994: 345).

Conclusion: Modern Science and European Colonialism: A Conversation with J. P. S. Uberoi and Bruno Latour

1 Uberoi was not the only social scientist in the non-West investigating the relationship between science and culture at this time. For example, in India, parallel to Uberoi's study, Claude Alvares (1980) and Ashish

Nandy (1995 [1980]) published their own independent analyses of the relationship between science and society in the context of a non-western society, and these studies are different from each other in their approach.

2 *Swaraj*, literally meaning self-rule, was, understandably, a central element of anti-colonial struggles in India. However, as the debates over independence from the colonial rule show, the term had much broader and also shifting implications; it included a concern with progressive transformation of the colonized societies. For example, Mahatma Gandhi, in his book *Hind Swaraj* (Gandhi 1938 [1910]) that is presented as a dialogue between an editor and a reader, categorically stated that *swaraj* cannot simply entail freedom from British colonial rule; it had to include broader transformation of people and society and as such should include, for example, belief in non-violence and *swadeshi* (indigenous production).

3 In fact, Ashis Nandy, another important proponent of alternative sciences, also "does not throw away modern science or pursuit for universal knowledge" (Prasad 2006: 220). Nandy shows how non-western/ Hindu social/psychological "habitats" of Indian scientists (Jagdish Chandra Bose and Srinivas Ramunajan) were in tension with the culture of modern science, but also productive sources of creativity for these scientists, resulting in the birth of alternative sciences (Nandy 1995 [1980]).

4 Uberoi argues that Zwingli's importance in the making of European modernity "is dismissed very lightly by the sociologist [Max] Weber" (Uberoi 1978: 30). Thus, by highlighting the role of Zwingli, he also presents a counter-narrative to Weber's claim of the birth of capitalism and with it that of European modernity with Calvinism (see M. Weber 2010).

5 According to Priya Satia, "E. P. Thompson's framing of British social history in the era of Indian national-

ism, as a kind of escape from decolonization, depended on denial of precisely those transnational bonds" that were forged through European colonialism (Satia 2020: 265).

6 Seeley does not erase the history of British/European imperialism. Indeed, his book is an attempt to historically situate the "expansion of England." Rather, he presents colonization of, for example, the Americas and also India as contingent effects of confrontations between the British and the French (Seeley 1884). Thus colonialism becomes another European affair that simultaneously uses and erases the non-western "others." Seeley's "book was an immense success, selling 80,000 copies within two years." It was "claimed that . . . 'since Sieyes no pamphlet has ever had such immediate and wide-reaching influence.'" https:// www.wikiwand.com/en/The_Expansion_of_England, accessed February 17, 2022.

7 Latour, for example, delves into the etymology of the term "social" to present another genealogy of sociology. He writes: "What I want to do is to redefine the notion of social by going back to its *original meaning* and making it able to trace connections again" (Latour 2005: 1; emphasis added).

References

Abraham, I. (1998). *The Making of the Indian Atomic Bomb: Science, Secrecy and the Postcolonial State.* Delhi: Zed Books.

Abraham, I. (2000). Landscape and Postcolonial Science. *Contributions to Indian Sociology* 34(2): 163–87.

Abraham, I. (2006). The Contradictory Spaces of Postcolonial Techno-science. *Economic and Political Weekly* 41(3): 210–17.

Abrami, R. M., Kirby, W. C., and McFarlan, F. W. (2014). Why China Can't Innovate. *Harvard Business Review* 92(3): 107–11.

Adas, M. (1989). *Machines as the Measure of Men: Science, Technology, and Ideologies of Western Dominance.* Ithaca, NY: Cornell University Press.

Althusser, L. (1971). *Lenin and Philosophy and Other Essays.* New York: Monthly Review Press.

Alvares, C. (1980). *Homo Faber: Technology and Culture in India, China and the West from 1500 to the Present Day.* Boston: Martinus Nijhoff Publishers.

Amin, S. (1995). *Event, Metaphor, Memory: Chauri Chaura, 1922–1992.* Delhi: Oxford University Press.

Anderson, B. (1991). *Imagined Communities: Reflections on the Origins and Spread of Nationalism.* New York: Verso.

Anderson, W. (2002). Postcolonial Technoscience. *Social Studies of Science* 32(5): 643–58.

Anderson, W. (2006). *The Cultivation of Whiteness: Science, Health, and Racial Destiny in Australia.* Durham, NC: Duke University Press.

Anderson, W. (2008). *The Collectors of Lost Souls: Turning Kuru Scientists into Whitemen.* Baltimore, MD: Johns Hopkins University Press.

Anderson, W. (2009). From Subjugated Knowledge to Conjugated Subjects: Science and Globalisation, or Postcolonial Studies of Science? *Postcolonial Studies* 12(4): 389–400.

Anderson, W. (2012). Asia as Method in Science and Technology Studies. *East Asian Science, Technology and Society: An International Journal* 6(4): 445–51.

Anderson, W. (2018). Remembering the Spread of Western Science. *Historical Records of Australian Science* 29(2): 73–81.

Anderson, W. (2020a). Finding Decolonial Metaphors in Postcolonial Histories. *History and Theory* 59(3): 430–8.

Anderson, W. (2020b). STS with East Asian Characteristics? *East Asian Science, Technology and Society: An International Journal* 14(1): 163–8.

Anderson, W. and Adams, V. (2008). Pramoedya's Chickens: Postcolonial Studies of Technoscience, in E. Hackett, O. Amsterdamska, M. Lynch, and J. Wajcman (eds), *The Handbook of Science and Technology Studies.* Cambridge, MA: MIT Press, pp. 181–204.

Anderson, W. and Prasad, A. (2017). "Things Do Look Different from Here, on the Borderlands": An Interview with Warwick Anderson. *Science, Technology & Society* 22(1): 135–43.

Arendt, H. (1945). Imperialism, Nationalism, Chauvinism. *Review of Politics* 7: 441–63.

Arnold, D. (1993). *Colonizing the Body: State Medicine*

and Epidemic Disease in Nineteenth-Century India. Delhi: Oxford University Press.

Arnold, D. (2013). Nehruvian Science and Postcolonial India. *Isis* 104: 360–70.

Asad, T. (ed.) (1973). *Anthropology and the Colonial Encounter.* New York: Humanity Books.

Bajaj, S. S. and Stanford, F. C. (2021). Beyond Tuskegee: Vaccine Distrust and Everyday Racism. *New England Journal of Medicine* 384(5): e12.

Banjerjee, D. (2020). Fantasies of Control: The Colonial Character of the Modi Government's Actions during the Pandemic. *The Caravan*, 10–15. Retrieved from: https://caravanmagazine.in/author/34837

Basalla, G. (1967). The Spread of Western Science. *Science* 156: 611–22.

Bauman, Z. (1993). *Modernity and Ambivalence.* Malden, MA: Polity.

Benton, L. (2010). *A Search for Sovereignty: Law and Geography in European Empires, 1400–1900.* New York: Cambridge University Press.

Bhabha, H. (1994). *The Location of Culture.* NewYork: Routledge.

Biagioli, M. (2006). Patent Republic: Representing Inventions, Constructing Rights and Authors. *Social Research: An International Quarterly* 73(4): 1129–72.

Binagwaho, A., Frisch, M., Ntawukuriyayo, J. T., and Hirschhorn, L. (2020). Changing the COVID-19 Narrative in Africa: Using an Implementation Research Lens to Understand Success and Plan for Challenges Ahead. *Annals of Global Health* 86(1): 1–5.

Bloom, J., Chan, Y. A., Baric, R., et al. (2021). Investigate the Origins of COVID-19. *Science* 372(6543): 694.

Bloor, D. (1991). *Knowledge and Social Imagery.* Chicago: University of Chicago Press.

Bourdieu, P. (1977). *Outline of a Theory of Practice.* New York: Cambridge University Press.

Boustany, Jr, C. W. and Friedberg, A. L. (2019). *Answering*

China's Economic Challenge: Preserving Power, Enhancing Prosperity. Retrieved from https://www. nbr.org/wp-content/uploads/pdfs/publications/special_ report_76_answering_chinas_economic_challenge.pdf

Bowen, H. V. (2006). *The Business of Empire: The East India Company and Imperial Britain, 1756–1833.* New York: Cambridge University Press.

Boxer, C. R. (1960). The *Carreira da India*, 1650–1750. *The Mariner's Mirror* 46(1): 35–54.

Broniatowski, D., Jamison, A., Johnson, N., et al. (2020). Facebook Pages, the "Disneyland" Measles Outbreak, and Promotion of Vaccine Refusal as a Civil Right, 2009–2019. *American Journal of Public Health* 110: S312–S318.

Burton, A. (1994). *Burdens of History: British Feminists, Indian Women, and Imperial Culture, 1865–1915.* Chapel Hill, NC: University of North Carolina Press.

Butler, J. (1990). *Gender Trouble.* New York: Routledge.

Butterfield, H. (1957). *The Origins of Modern Science.* New York: Free Press.

Butterfield, H. (1965 [1931]). *The Whig Interpretation of History.* New York: W. W. Norton.

Butterfield, H. (1981). *The Origins of History*, ed. A. Watson. London: Eyre Methuen.

Calisher, C., Carroll, D., Colwell, R., et al. (2020). Statement in Support of the Scientists, Public Health Professionals, and Medical Professionals of China Combatting COVID-19. *Lancet* 395: E42–E43.

Callahan, W. (2004). National Insecurities: Humiliation, Salvation, and Chinese Nationalism. *Alternatives* 29: 199–218.

Callon, M. (1986). Some Elements of a Sociology of Translation: Domestication of Scallops and the Fishermen of St Brieuc Bay, in J. Law (ed.), *Power, Action and Belief: A New Sociology of Knowledge.* London: Routledge & Kegan Paul, pp. 196–233.

Callon, M. and Law, J. (1995). Agency and the Hybrid "Collectif." *South Atlantic Quarterly* 94: 481–507.

Chakrabarty, D. (2000). *Provincializing Europe: Postcolonial Thought and Historical Difference.* Princeton, NJ: Princeton University Press.

Chambers, D. W. (1987). Period and Process in Colonial and National Science, in N. Reingold and M. Rothenberg (eds), *Scientific Colonialism: A Cross-Cultural Comparison.* Washington, DC: Smithsonian Institution Press.

Charlier, P. and Varison, L. (2020). Is COVID-19 Being Used as a Weapon against Indigenous Peoples in Brazil? *Lancet* 396(10257): 1069–70.

Chow, R. (1993). *Writing Diaspora: Tactics of Intervention in Contemporary Cultural Studies.* Bloomington, IN: Indiana University Press.

Chowkwanyun, M. and Reed, A. (2020). Racial Health Disparities and Covid-19: Caution and Context. *New England Journal of Medicine* 383: 201–3.

Cohen, H. F. (1994). *The Scientific Revolution: A Historiographical Inquiry.* Chicago: University of Chicago Press.

Cohen, I. B. (1976). The Eighteenth-Century Origins of the Concept of Scientific Revolution. *Journal of History of Ideas* 37(2): 257–88.

Cohen, I. B. (1985). *Revolution in Science.* Cambridge, MA: Harvard University Press.

Cohen, L. (1994). Whodunit? Violence and the Myth of Fingerprints: Comment on Harding. *Configurations* 2(2): 343–7.

Collins, P. H. (1996). What's in a Name? Womanism, Black Feminism, and Beyond. *Journal of Black Studies and Research* 26(1): 9–17.

Collins, P. H. (1998). It's All in the Family: Intersections of Gender, Race, and Nation. *Hypatia* 13(3): 62–82.

Collins-Dexter, B. (2020). Canaries in the Coal Mine: COVID-19 Misinformation and Black Communities.

Retrieved from https://shorensteincenter.org/wp-content/uploads/2020/06/Canaries-in-the-Coal-Mine-Shorenstein-Center-June-2020.pdf

Coninck, D. D., Frissen, T., Matthijs, K., et al. (2021). Beliefs in Conspiracy Theories and Misinformation About COVID-19: Comparative Perspectives on the Role of Anxiety, Depression and Exposure to and Trust in Information Sources. *Frontiers in Psychology*. Retrieved from: https://doi.org/10.3389/fpsyg.2021.646394

Cunningham, A. and Williams, P. (1993). De-centering the "Big Picture": *The Origins of Modern Science* and the Modern Origins of Science. *British Journal for the History of Science* 26(4): 407–32.

Dalrymple, W. (2019). *The Anarchy: The Relentless Rise of the East India Company*. New York: Bloomsbury Publishing.

Das, V. (1984). *Critical Events: An Anthropological Perspective of Contemporary India*. Delhi: Oxford University Press.

Daston, L. (1991). The Ideal and Reality of Republic of Letters in the Enlightenment. *Science in Context* 4(2): 367–86.

Daston, L. (2015). History of Science, in J. D. Wright (ed.), *International Encyclopedia of the Social & Behavioral Sciences*, 2nd edn. Amsterdam: Elsevier, vol. 11, pp. 241–7.

Davis, A. (1983). *Women, Race, and Class*. New York: Vintage Books.

Dear, P. (2005). What Is the History of Science the History Of? Early Modern Roots of the Ideology of Modern Science. *Isis* 96(3): 390–406.

Dear, P. (2009). The History of Science and the History of the Sciences. *Isis* 100(1): 89–93.

Dear, P. (2012). Science Is Dead; Long Live Science. *Osiris* 27(1): 37–55.

Derrida, J. (1978). *Writing and Difference*. Chicago: University of Chicago Press.

Derrida, J. (1998). *Of Grammatology*, trans. G. C. Spivak. Baltimore, MD: Johns Hopkins University Press.

Dirks, N. (2006). *The Scandal of Empire: India and the Creation of Imperial Britain*. Cambridge, MA: Harvard University Press.

Dredze, M., Broniatowski, D., Smith, M., and Hilyard, K. (2016). Understanding Vaccine Refusal: Why We Need Social Media Now. *American Journal of Preventive Medicine*, 50(4): 550–2.

Du Bois, W. E. B. (1997 [1903]). *The Souls of Black Folk*. New York: Macmillan Publishing.

Du Bois, W. E. B. (2014 [1920]). *Darkwater: Voices from Within the Veil*. Mansfield Centre, CT: Martino Publishing.

Elkana, Y. (1987). Alexandre Koyre: Between the History of Ideas and Sociology of Knowledge. *History and Technology* 4: 115–48.

Elshakry, M. (2008). Knowledge in Motion: The Cultural Politics of Modern Science Translations in Arabic. *Isis* 99: 701–30.

Elshakry, M. (2010). When Science Became Western: Historiographical Reflections. *Isis* 101: 98–109.

Fabian, J. (2006). The Other Revisited: Critical Afterthoughts. *Anthropological Theory* 6(2): 139–52.

Falk, S. (2014). The Scholar as Craftsman: Derek De Solla Price and the Reconstruction of a Medieval Instrument. *Notes and Records: The Royal Society Journal of the History of Science* 2014(68): 111–34.

Fan, F.-t. (2007a). Redrawing the Map of Science in Twentieth-Century China. *Isis* 98(3): 524–38.

Fan, F.-t. (2007b). Science in Cultural Borderlands: Methodological Reflections on the Study of Science, European Imperialism, and Cultural Encounter. *East Asian Science, Technology and Society: An International Journal* 1: 213–31.

Fan, F.-t. (2016). Modernity, Region, and Technoscience:

One Small Cheer for Asia as Method. *Cultural Sociology* 10(3): 352–68.

Farquhar, J. (2002). *Appetitites: Food and Sex in Post-socialist China*. Durham, NC: Duke University Press.

Farquhar, J. and Zhang, Q. (2012). *Ten Thousand Things: Nurturing Life in Contemporary Beijing*. New York: Zone Books.

Ferguson, N. (2002). *Empire: The Rise and Demise of the British World Order and the Lessons for Global Power*. New York: Basic Books.

Forman, P. (2012). On the Historical Forms of Knowledge Production and Curation: Modernity Entailed Disciplinarity, Postmodernity Entails Antidisciplinarity. *Osiris* 27(1): 56–97.

Foucault, M. (1977a). Nietzsche, Genealogy, History, in D. Bouchard (ed.), *Language, Counter-Memory, Practice: Selected Essays and Interviews*. Ithaca, NY: Cornell University Press, pp. 139–64.

Foucault, M. (1977b). *What Is an Author?* Ithaca, NY: Cornell University Press.

Foucault, M. (2011). *The Government of Self and Others: Lectures at Collège de France, 1982–83*, trans. G. Burchell. New York: Palgrave Macmillan.

Franklin, S. (1995). Science as Culture, Cultures of Science. *Annual Review of Anthropology* 24(1): 163–84.

Fujimura, J. and Luce, H. (1998). Authorizing Knowledge in Science and Anthropology. *American Anthropologist* 100(2): 347–60.

Gandhi, M. K. (1938 [1910]). *Hind Swaraj or Indian Home Rule*. Ahmedabad: Navajivan Publishing House.

Garfield, E. (1985). The Life and Career of George Sarton: The Father of the History of Science. *The History of the Behavioral Sciences* 21(2): 107–17.

Garland, D. (2014). What Is a "History of the Present"? On Foucault's Genealogies and Their Critical Preconditions. *Punishment & Society* 16(4): 365–84.

Ghosh, A. (1992). *In an Antique Land*. New Delhi: Permanent Black.

Ghosh, A. (2008). The Role of Rumour in History Writing. *History Compass* 6(5): 1235–43.

Golinski, J. (2012). Is It Time to Forget Science? Reflections on Singular Science and Its History. *Osiris* 27(1): 19–36.

Gross, P. and Levitt, N. (1994). *Higher Superstition: The Academic Left and Its Quarrels with Science*. Baltimore, MD: Johns Hopkins University Press.

Guha, R. (1983). *Elementary Aspects of Peasant Insurgency*. Delhi: Oxford University Press.

Habib, I. and Raina, D. (eds) (1999). *Situating the History of Science: Dialogues with Joseph Needham*. Delhi: Oxford University Press.

Hacking, I. (2004). *Historical Ontology*. Cambridge, MA: Harvard University Press.

Hall, S. (1996). When Was "the Postcolonial"? Thinking at the Limit, in I. Chambers and L. Curti (eds), *The Post-Colonial Question*. New York: Routledge, pp. 242–60.

Haraway, D. (1991). *Simians, Cyborgs, and Women: The Reinvention of Nature*. New York: Routledge.

Haraway, D. (1997). *Modest Witness@Second Millennium. FemaleMan© Meets OncoMouse™: Feminism and Technoscience*. New York: Routledge.

Harding, S. (1991). *Whose Science? Whose Knowledge? Thinking from Women's Lives*. Ithaca, NY: Cornell University Press.

Harding, S. (1994). Is Science Multicultural? Challenges, Resources, Opportunities, Uncertainties. *Configurations* 2(2): 301–30.

Harding, S. (1998). *Is Science Multicultural? Postcolonialism, Feminisms and Epistemologies*. Bloomington, IN: Indiana University Press.

Harding, S. (2011a). Beyond Postcolonial Theory: Two Undertheorized Perspectives on Science and Technology, in S. Harding (ed.), *The Postcolonial Science and Technology Studies Reader*. Durham, NC: Duke University Press, pp. 1–31.

Harding, S. (ed.) (2011b). *The Postcolonial Science and Technology Studies Reader*. Durham, NC: Duke University Press.

Hart, R. (1999). Beyond Science and Civilization: A Post-Needham Critique. *East Asian Science, Technology and Society: An International Journal* 16: 88–114.

Headrick, D. (1988). *The Tentacles of Progress: Technology Transfer in the Age of Imperialism 1850–1940*. New York: Oxford University Press.

Hecht, G. (2002). Rupture-Talk in the Nuclear Age: Conjugating Colonial Power in Africa. *Social Studies of Science* 32(5–6): 691–727.

Hofmänner, A. (2015). Science Studies Elsewhere: The Experimental Life and the Other Within. *Social Epistemology* 30(2): 186–212.

Home, R. W. (2002). The Royal Society and the Empire: The Colonial and Commonwealth Fellowship, Part 1. 1731–1847. *Notes and Records of the Royal Society of London* 56(3): 303–32.

Hooper, M. W., Napoles, A. M., and Perez-Stable, E. (2020). COVID-19 and Racial/Ethnic Disparities. *JAMA* 323(24): 2466–7.

Hotez, P. J. (2021). *Preventing the Next Pandemic: Vaccine Diplomacy in a Time of Anti-Science*. Philadelphia: Johns Hopkins University Press.

Hsia, F. and Schäfer, D. (2019). History of Science, Technology, and Medicine: A Second Look at Joseph Needham. *Isis* 110(1): 94–9.

Hswen, Y., Xu, X., Hing, A., Hawkins, J., Brownstein, J., and Gee, G. (2021). Association of "#covid19" versus "#chinesevirus" with Anti-Asian Sentiments on Twitter: March 9–23, 2020. *American Journal of Public Health* 111(5): 956–64.

Hughes, D. M. (2006). Hydrology of Hope: Farm Dams,

Conservation, and Whiteness in Zimbabwe. *American Ethnologist* 33(2): 269–87.

Inden, R. (1986). Orientalist Constructions of India. *Modern Asian Studies* 20(3): 401–46.

Irani, L., Vertesi, J., Dourish, P., Kavita, P., and Grinter, R. (2010). Postcolonial Computing: A Lens on Design and Development. Paper presented at the SIGCHI Conference on Human Factors in Computing Systems, Atlanta, GA.

Irwin, A. and Wynne, B. (1996). Introduction, in A. Irwin and B. Wynne (eds), *Misunderstanding Science? The Public Reconstruction of Science and Technology*. New York: Cambridge University Press, pp. 1–18.

Islam, M. S., Sarkar, T., Khan, S. H., et al. (2020). COVID-19-Related Infodemic and Its Impact on Public Health: A Global Social Media Analysis. *American Journal of Tropical Medicine and Hygiene* 103(4): 1621–9.

Jameson, F. (1988). *Cognitive Mapping*. Urbana: University of Illinois Press.

Jameson, F. (1992). *Postmodernism, or, The Cultural Logic of Late Capitalism*. Durham, NC: Duke University Press.

Jasanoff, S. (1987). Contested Boundaries in Policy-Relevant Science. *Social Studies of Science* 17(2): 195–230.

Jasanoff, S. (2000). The "Science Wars" and American Politics, in M. Dierkes and C. von Grote (eds), *Between Understanding and Trust: The Public, Science and Technology*. Amsterdam: Harwood, pp. 39–60.

Jasanoff, S. (2005). *Designs on Nature: Science and Democracy in Europe and the United States*. Princeton, NJ: Princeton University Press.

Jasanoff, S. (2011). Making the Facts of Life, in S. Jasanoff (ed.), *Reframing Rights: Bioconstitutionalism in the Genetic Age*. Cambridge, MA: MIT Press, pp. 59–83.

Jensen, C. B. and Blok, A. (2013). Techno-animism in Japan: Shinto Cosmograms, Actor–Network Theory, and the Enabling Powers of Non-human Agency. *Theory, Culture & Society* 30(2): 84–115.

Johnson, J. and Latour, B. (1988). Mixing Humans with Non-humans: Sociology of a Door Opener. *Social Problems* 35: 298–310.

Kapila, S. (2010). The Enchantment of Science in India. *Isis* 101(1): 120–32.

Kennedy, J. (2019). Populist Politics and Vaccine Hesitancy in Western Europe: An Analysis of National-Level Data. *European Journal of Public Health* 29(3): 512–16.

Kissinger, H. (1977). *American Foreign Policy*. New York: W. W. Norton.

Kleinman, D. L. (1998). Beyond the Science Wars: Contemplating the Democratization of Science. *Politics and the Life Sciences* 17(2): 133–45.

Koch, A., Brierley, C., Maslin, M., and Lewis, S. (2019). Earth System Impacts of the European Arrival and Great Dying in the Americas after 1492. *Quarterly Science Reviews* 207: 13–36.

Kohler, R. and Olesko, K. (2012). Introduction: Clio Meets Science. *Osiris* 27(1): 1–16.

Koyre, A. (1943). Galileo and the Scientific Revolution of the Seventeenth Century. *Philosophical Review* 52(4): 333–48.

Krishna, V. V. (1992). The Colonial Model and the Emergence of National Science in India, 1876–1920, in P. Petitjean, C. Jami, and A. M. Moulin (eds), *Science and Empires: Historical Studies about Scientific Development and European Expansion*. The Hague: Kluwer Academic Publishers, pp. 57–72.

Kuhn, T. (1970). *The Structure of Scientific Revolutions*. Chicago: University of Chicago Press.

Laet, M. de and Mol, A. (2000). The Zimbabwe Bush Pump: Mechanics of a Fluid Technology. *Social Studies of Science* 30(2): 225–63.

Lake, M. (2004). The White Man under Siege: New Histories of Race in the Nineteenth Century and the

Advent of White Australia. *History Workshop Journal* 58: 41–62.

Lam, T. (2011). *A Passion for Facts: Social Surveys and the Construction of the Chinese Nation-State, 1900–1949.* Berkeley: University of California Press.

Latour, B. (1987). *Science in Action: How to Follow Scientists and Engineers through Society.* Cambridge, MA: Harvard University Press.

Latour, B. (1988). *The Pasteurization of France*, trans A. Sheridan. Cambridge, MA: Harvard University Press.

Latour, B. (1993). *We Have Never Been Modern*, trans. C. Porter. Cambridge, MA: Harvard University Press.

Latour, B. (1999). *Pandora's Hope: Essay on the Reality of Science Studies.* Cambridge, MA: Harvard University Press.

Latour, B. (2005). *Reassembling the Social: An Introduction to Actor–Network Theory.* Oxford: Oxford University Press.

Law, J. (1986). On the Methods of Long-Distance Control: Vessels, Navigation and the Portuguese Route to India, in M. Callon and J. Law (eds), *Power, Action and Belief: A New Sociology of Knowledge.* New York: Routledge, pp. 234–63.

Law, J. (1987). On the Social Explanation of Technical Change: The Case of the Portuguese Maritime Expansion. *Technology and Culture* 28(2): 227–52.

Law, J. (1992). Notes on the Theory of the Actor–Network: Ordering, Strategy and Heterogeneity. *Systems Practice* 5(4): 379–93.

Law, J. and Lin, W.-y. (2017). Provincializing STS: Postcoloniality, Symmetry, and Method. *East Asian Science, Technology and Society: An International Journal* 11: 211–27.

Lederberg, J. (2000). Infectious History. *Science* 288(5464): 287–93.

Lerner, L. (2000). Good and Bad Science in US Schools. *Nature* 407: 287–90.

Lévi-Strauss, C. (1966). *The Savage Mind*. Chicago: University of Chicago Press.

Li, H. O.-Y., Bailey, A., Huynh, D., and Chan, J. (2020). YouTube as a Source of Information on COVID-19: A Pandemic of Misinformation? *BMJ Global Health* 5(5): 1–8. Retrieved from: https://gh.bmj.com/content/bmjgh/5/5/e002604.full.pdf?with-ds=yes

Libin, X. and Patapan, H. (2020). Schmitt Fever: The Use and Abuse of Carl Schmitt in Contemporary China. *International Journal of Constitutional Law* 18(1): 130–46.

Lim, D. (2019). The US, China and "Technology War." *Global Asia* 14(1): 8–13.

Lin, W.-y. and Law, J. (2014). A Correlative STS: Lessons from a Chinese Medical Practice. *Social Studies of Science* 44(6): 801–24.

Lin, W.-y. and Law, J. (2015). We Have Never Been Latecomers!? Making Knowledge Spaces for East Asian Technosocial Practices. *East Asian Science, Technology and Society: An International Journal* 9: 117–26.

Lin, W.-y. and Law, J. (2019). Where Is East Asia in STS? *East Asian Science, Technology and Society: An International Journal* 13: 115–36.

Lindberg, D. (1990). Conceptions of the Scientific Revolution from Bacon to Butterfield: A Preliminary Sketch, in D. Lindberg and R. Westman (eds), *Reappraisals of the Scientific Revolution*. New York: Cambridge University Press, pp. 1–26.

Lindberg, D. and Westman, R. (eds). (1990). *Reappraisals of the Scientific Revolution*. New York: Cambridge University Press.

Lo Presti, R. (2014). History of Science: The First Scientist. *Nature* 512: 250–1.

Lynch, K. (1960). *The Image of the City*. Boston: MIT Press.

Lynch, M. (1993). *Scientific Practice and Ordinary Action:*

Ethnomethodology and Social Studies of Science. New York: Cambridge University Press.

Lynch, M. (2020). We Have Never Been Anti-Science: Reflections on Science Wars and Post-Truth. *Engaging Science, Technology and Society* 6: 49–57.

MacLeod, R. (1987). *On Visiting the Moving Metropolis: Reflections on the Architecture of Imperial Science*. Washington, DC: Smithsonian Institution Press.

Marbot, O. (2020). Coronavirus: Unpacking the Theories Behind Africa's Low Infection Rate. Retrieved from: https://www.theafricareport.com/27470/coronavirus-unpacking-the-theories-behind-africas-low-infection-rate/

Mavhunga, C. C. (2014). *Transient Workspaces: Technologies of Everyday Innovation in Zimbabwe*. Cambridge, MA: MIT Press.

Mavhunga, C. C. (2017). The Language of Science, Technology, and Innovation: A *Chimurenga* Way of Seeing from *Dzimbahwe*, in C. C. Mavhunga (ed.), *What Do Science, Technology, and Innovation Mean from Africa?* Cambridge, MA: MIT Press, pp. 45–77.

Mbembe, A. (2017). *Critique of Black Reason*, trans. L. Dubois. Durham, NC: Duke University Press.

McCright, A. M. and Dunlap, R. E. (2011). Cool Dudes: The Denial of Climate Change among Conservative White Males in the United States. *Global Environmental Change* 21(4): 1163–72.

McLaren, J. (2021). Racial Disparity in COVID-19 Deaths: Seeking Economic Roots with Census Data. *B. E. Journal of Economic Analysis & Policy*. Retrieved from: https://doi.org/10.1515/bejeap-2020-0371

McNeil, M. (2005). Postcolonial Technoscience. *Science as Culture* 14(2): 105–200.

Mehta, U. (1999). *Liberalism and Empire: A Study in Nineteenth-Century British Liberal Thought*. Chicago: University of Chicago Press.

Meighoo, S. (2008). Derrida's Chinese Prejudice. *Cultural Critique* 68(Winter):, 163–209.

Mikovits, J. and Heckenlively, K. (2020). *Plague of Corruption: Restoring Faith in the Promise of Science.* New York: Skyhorse Publishing.

Milhaupt, C. and Callahan, M. (2021). The Rule of Law in the US–China Tech War. SSRN: Retrieved from https://ssrn.com/abstract=3840584

Mirzoeff, N. (2018). It's Not the Anthropocene: It's the White Supremacy Scene; or, The Geographical Color Line, in R. Grusin (ed.), *After Extinction.* Minneapolis, MN: University of Minnesota Press, pp. 123–49.

Mizuno, H. (2009). *Science for the Empire: Scientific Nationalism in Modern Japan.* Stanford, CA: Stanford University Press.

Mlambo, A. (2005). "Land Grab" or "Taking Back Stolen Land": The Fast Track Land Reform Process in Zimbabwe in Historical Perspective. *History Compass* 3: 1–21.

Mohanty, C. (1988). Under Western Eyes: Feminist Scholarship and Colonial Discourses. *Feminist Review* 30(1): 61–88.

Mol, A. (2010). Actor–Network Theory: Sensitive Terms and Enduring Tensions. *Kölner Zeitschrift für Soziologie und Sozialpsychologie* 50(1): 253–69.

Moran, P. (2020). Social Media: A Pandemic of Misinformation. *American Journal of Medicine,* 133(11): 1247–8.

Morgan, P. (1990). *Rural Water Supplies and Sanitation: A Text from Zimbabwe's Blair Research Institute.* London: Macmillan.

Morita, A. (2014). The Ethnographic Machine: Experimenting with Context and Comparison. *Science, Technology & Human Values* 39(2): 214–35.

Mouffe, C. (1997). Carl Schmitt and the Paradox of Liberal Democracy. *Canadian Journal of Law and Jurisprudence* 10(1): 21–34.

Moya, S. (2000). *Land Reform under Structural Adjustment in Zimbabwe.* Stockholm: Nordic Africa Institute.

Moya, S. (2004). *The Politics of Land Distribution and Race Relations in Southern Africa*. Geneva, Switzerland: United Nations Research Institute for Social Development.

Munro, W. (1941). Clio and Her Cousins: Some Reflection upon the Place of History among the Social Sciences. *Pacific Historical Review* 10(4): 403–10.

Nair, J. (1992). Uncovering the Zenana: Visions of Indian Womanhood in Englishwomen's Writings, 1813–1940, in C. Johnson-Odim and M. Strobel (eds), *Expanding the Boundaries of Women's History: Essays on Women in the Third World*. Bloomington, IN: Indiana University Press.

Nanda, M. (1997). The Science Wars in India. *Dissent* 44(1): 78–83.

Nandy, A. (1995 [1980]). *Alternative Sciences: Creativity and Authenticity in Two Indian Scientists*. Delhi: Oxford University Press.

Narayan, U. (1997). *Dislocating Cultures: Identities, Traditions, and Third-World Feminism*. New York: Routledge.

Nash, J. (2019). *Black Feminism Reimagined: After Intersectionality*. Durham, NC: Duke University Press.

Needham, J. (1969a). *The Grand Titration: Science and Society in East and West*. London: George Allen & Unwin.

Needham, J. (1969b). *Within the Four Seas: The Dialogue of East and West*. Buffalo, NY: University of Toronto Press.

Needham, J. (2004). *Science and Civilization in China* (Vol. 7: Part II), ed. K. G. Robinson. New York: Cambridge University Press.

Nyambi, O. (2017). "The Blair that I Know Is a Toilet": Political Nicknames and Hegemonic Control in Post-2000 Zimbabwe. *African Identities* 15(2): 143–58.

Osler, M. (2000a). The Canonical Imperative: Rethinking the Scientific Revolution, in M. Osler (ed.), *Rethinking the Scientific Revolution*. New York: Cambridge University Press, pp. 3–24.

Osler, M. (ed.) (2000b). *Rethinking the Scientific Revolution*. New York: Cambridge University Press.

Pacey, A. (1992). *The Maze of Ingenuity: Ideas and Idealism in the Development of Technology*. Cambridge, MA: MIT Press.

Pearson, C. (1893). *National Life and Character: A Forecast*. New York: Macmillan.

Peters, M. A., McLaren, P., and Jandric, P. (2020). A Viral Theory of Post-Truth. *Educational Philosophy and Theory*. Retrieved from: https://www.tandfonline.com/doi/full/10.1080/00131857.2020.1750090

Philip, K. (2004). *Civilizing Natures: Race, Resources, and Modernity in Colonial South India*. New Brunswick, NJ: Rutgers University Press.

Philip, K., Irani, L., and Dourish, P. (2012). Postcolonial Computing: A Tactical Survey. *Science, Technology & Human Values* 37(1): 3–29.

Pickering, A. (1995). *The Mangle of Practice: Time, Agency, and Science*. Chicago: Chicago University Press.

Pocock, J. G. A. (1997). What Do We Mean by Europe? *The Wilson Quarterly* 21(1): 12–29.

Polanyi, M. (1951). *The Logic of Liberty: Reflections and Rejoinders*. London: Routledge & Kegan Paul.

Polanyi, M. (1962). The Republic of Science: Its Political and Economic Theory. *Minerva* 1 (Autumn): 1–20.

Polanyi, M., Ziman, J., and Fuller, S. (2000). The Republic of Science: Its Political and Economic Theory. *Minerva* 38(1): 1–32.

Prakash, G. (1992). Postcolonial Criticism and Indian Historiography. *Social Text* 31/32: 8–19.

Prakash, G. (1999). *Another Reason: Science and the Imagination of Modern India*. Princeton, NJ: Princeton University Press.

Prasad, A. (2005). Scientific Culture in the Other Theatre of Modern Science: An Analysis of the Culture of Magnetic

Resonance Imaging (MRI) Reasearch in India. *Social Studies of Science* 30(3): 463–89.

Prasad, A. (2006). Beyond Modern versus Alternative Science Debate: Analysis of Magnetic Resonance Imaging Research. *Economic & Political Weekly* 41(3): 219–27.

Prasad, A. (2014). *Imperial Technoscience: Transnational Histories of MRI in the United States, Britain, and India.* Cambridge, MA: MIT Press.

Prasad, A. (2016). Discursive Contextures of Science: Euro/West-Centrism and Science and Technology Studies. *Engaging Science, Technology and Society* 2: 193–207.

Prasad, A. (2017). Introduction: Global Assemblages of Technoscience. *Science, Technology & Society* 22(1): 1–5.

Prasad, A. (2018). Taj Mahal, Circulations of Science, and (Post) Colonial Present. *History and Technology* 34(1): 51–60.

Prasad, A. (2019). Burdens of the Scientific Revolution: Euro/West-Centrism, Black Boxed Machines, and the (Post) Colonial Present. *Technology and Culture* 60(4): 1059–82.

Prasad, A. (2022). Anti-Science Conspiracies: COVID-19, Post-truth, and Decolonial Science Studies (STS). *Science, Technology & Society* 27(1): 88–112.

Quijano, A. (2000). Coloniality of Power and Eurocentrism in Latin America. *International Sociology* 15(2): 215–32.

Raina, D. (1995). Homage to an Honorary Taoist. *Economic & Political Weekly* 30(30): 1904–6.

Raina, D. (1996). Reconfiguring the Centre: The Structure of Scientific Exchanges between Colonial India and Europe. *Minerva* 34(2): 161–76.

Raina, D. (1999). Introduction, in S. I. Habib and D. Raina (eds), *Situating the History of Science: Dialogues with Joseph Needham.* New Delhi: Oxford University Press, pp. 1–15.

Raina, D. (2020). *Needham's Indian Network: The Search for a Home for the History of Science in India (1950–1970)*. New Delhi: Yoda Press.

Raj, K. (2007). *Relocating Modern Science: Circulation and the Construction of Knowledge in South Asia and Europe, 1650–1900*. New York: Palgrave Macmillan.

Raj, K. (2016). Rescuing Science from Civilisation: On Joseph Needham's "Asiatic Mode of (Knowledge) Production," in A. Bala and P. Duara (eds), *The Bright Dark Ages: Comparative and Connective Perspectives*. Leiden: Brill, pp. 255–80.

Redfield, P. (2002). The Half-Life of Empire in Outer Space. *Social Studies of Science* 32(5–6): 791–825.

Reilly, S. (2014). The Technology of Truth: Revisiting Areopagitica. *The International Journal of Technology, Knowledge and Society* 9(2): 139–44.

Relman, D., Hamburg, M., Choffnes, E., and Mack, A. (2009). *Microbial Evolution and Co-adaptation: A Tribute to the Life and Scientific Legacies of Joshua Lederberg*. Washington, DC: National Academies Press.

Reverby, S. (2021). Racism, Disease, and Vaccine Refusal: People of Color are Dying for Access of COVID-19 Vaccines. *PLos Biology* 19(3).

Robinson, A. (2002). *The Zimbabwe Experience: Lessons from a Review of 15 Years of the Zimbabwe Integrated Rural Water Supply and Sanitation Program*. Retrieved from https://documents1.worldbank.org/curated/en/90520146 8179070857/pdf/265490WSP0Field0Note0Zimbabwe 0experience.pdf

Rostow, W. W. (1990 [1960]). *The Stages of Economic Growth: A Non-communist Manifesto*. New York: Cambridge University Press.

Said, E. (1979). *Orientalism*. New York: Vintage.

Said, E. (1989). Representing the Colonized: Anthropology's Interlocutors. *Critical Inquiry* 15(2): 205–25.

Sakai, N. (2001). "You Asians": On the Historical Role of the West and Asia Binary. *South Atlantic Quarterly* 99(4): 789–817.

Sarton, G. (1916). The History of Science. *The Monist* 26: 321–65.

Sarton, G. (1924). The New Humanism. *Isis* 6: 9–42.

Sarton, G. (1953). Why Isis? *Isis* 44(3): 232–42.

Satia, P. (2020). *Time's Monster: History, Conscience and Britain's Empire*. London: Allen Lane.

Schmitt, C. (2006). *The Nomos in the International Law of the Jus Publicum Europaeum*, trans. G. L. Ulmen. Candor, NY: Telos Press Publishing.

Schmitt, C. (2007). *The Concept of the Political*, trans. G. Schwab. Chicago: University of Chicago Press.

Seeley, J. R. (1884). *The Expansion of England: Two Courses of Lectures*. Leipzig: Bernhard Tauchnitz.

Seth, S. (2009). Putting Knowledge in Its Place: Science, Colonialism, and the Postcolonial. *Postcolonial Studies* 12(4): 373–88.

Seth, S. (2017). Colonial History and Postcolonial Science Studies. *Radical History Review* 127(1): 63–85.

Shapin, S. (1996). *The Scientific Revolution*. Chicago: University of Chicago Press.

Shapin, S. (1998). Placing the View from Nowhere: Historical and Sociological Problems in the Location of Science. *Transactions of the Institute of British Geographers* 23(1): 5–12.

Sharma, P. and Anand, A. (2020). Indian Media Coverage of Nizamuddin Markaz Event during COVID-19 Pandemic. *Asian Politics & Policy* 12: 650–4.

Shim, D. (1998). From Yellow Peril through Model Minority to Renewed Yellow Peril. *Journal of Communication Inquiry* 22(4): 385–409.

Skocpol, T. (2014). Making Sense of the Past and Future Politics of Global Warming in the United States. Paper presented at the Max Weber Lecture Series.

Spivak, G. C. (1988). Can the Subaltern Speak? in C. Nelson

and L. Grossberg (eds), *Marxism and the Interpretation of Culture*. Urbana, IL: University of Illinois Press.

Spivak, G. C. (1998). Translator's Preface, in Jacques Derrida, *Of Grammatology*, trans. G. C. Spivak. Baltimore, MD: Johns Hopkins University Press, pp. ix–lxxxix.

Stern, P. (2009). History and Historiography of the English East India Company: Past, Present, and Future! *History Compass* 7(4): 1146–80.

Stingl, A. (2016). *The Digital Coloniality of Power*. Lanham, MA: Lexington Books.

Stoler, A. L. (2010a). *Along the Archival Grain: Epistemic Anxieties and Colonial Common Sense*. Princeton, NJ: Princeton University Press.

Stoler, A. L. (2010b). *Carnal Knowledge and Imperial Power: Race and the Intimate in Colonial Rule*. Berkeley: University of California Press.

Stoler, A. L. (2013). Introduction – "The Rot Remains": From Ruins to Ruination, in A. L. Stoler (ed.), *Imperial Debris: On Ruins and Ruination*. Durham, NC: Duke University Press, pp. 1–35.

Subrahmanyam, S. (1997a). *The Career and Legend of Vasco da Gama*. Cambridge, UK: Cambridge University Press.

Subrahmanyam, S. (1997b). Connected Histories: Notes Towards a Reconfiguration of Early Modern Eurasia. *Modern Asian Studies* 31(3): 735–62.

Subrahmanyam, S. (2005a). *Explorations in Connected History: Mughals and Franks*. Delhi: Oxford University Press.

Subrahmanyam, S. (2005b). *Explorations in Connected History: From the Tagus to the Ganges*. Delhi: Oxford University Press.

Subramaniam, B. (2000). Arachaic Modernities: Science, Secularism, and Religion in Modern India. *Social Text* 18(3): 67–86.

Sur, A. (1999). Aesthetics, Authority, and Control in an

Indian Laboratory: The Raman–Born Controversy on Lattice Dynamics. *Isis* 90(1): 25–49.

Sur, A. (2002). Scientism and Social Justice: Meghnad Saha's Critique of the State of Science in India. *Historical Studies in the Physical and Biological Sciences* 33(1): 87–105.

Suzuki, Y. (2018). The Good Farmer: Morality, Expertise, and Articulations of Whiteness in Zimbabwe. *Anthropogical Forum* 28(1): 74–88.

Taussig, M. (1987). *Shamanism, Colonialism, and the Wild Man: A Study of Terror and Healing.* Chicago: University of Chicago Press.

Thackray, A. (1984). Editorial: Sarton, Science, and History. *Isis* 75(1): 6–9.

Thompson, E. P. (1966). *The Making of the English Working Class.* New York: Vintage Books.

Toscano, A. (2008). Carl Schmitt in Beijing: Partisanship, Geopolitics and the Demolition of the Eurocentric World. *Postcolonial Studies* 11(4): 417–33.

Traweek, S. (1988). *Beamtimes and Lifetimes: The World of High Energy Physicists.* Cambridge, MA: Harvard University Press.

Uberoi, J. P. S. (1971). *Politics of Kula Ring: An Analysis of the Findings of Bronislaw Malinowski.* Manchester, UK: Manchester University Press.

Uberoi, J. P. S. (1978). *Science and Culture.* Delhi: Oxford University Press.

Uberoi, J. P. S. (1984). *The Other Mind of Europe: Goethe as a Scientist.* Delhi: Oxford University Press.

Verges, F. (2017). Racial Capitalocene, in G. T. Johnson and A. Lubin (eds), *Futures of Black Radicalism.* Brooklyn, NY: Verso, pp. 72–82.

Verne, J. and Verne, M. (2017). Introduction: The Indian Ocean as Aesthetic Space. *Comparative Studies of South Asia, Africa and the Middle East* 37(2): 314–20.

Verran, H. (2002). A Postcolonial Moment in Science

Studies: Alternative Firing Regimes of Environmental Scientists and Aboriginal Landowners. *Social Studies of Science* 32(5–6): 729–62.

Vrieze, J. de (2017). "Science Wars" Veteran Has a New Mission. *Science* 358(6360): 159–60.

Walsh, S. P. (1983). Community Participation in Zimbabwe. *Waterlines* 2(2): 14–16.

Wang, Z. (2014a). The Chinese Dream: Concept and Context. *Journal of Chinese Political Science* 19(1): 1–13.

Wang, Z. (2014b). *Never Forget National Humiliation: Historical Memory in Chinese Politics and Foreign Relations*. New York: Columbia University Press.

Weber, E. (1976). *Peasants into Frenchmen: The Modernization of Rural France*. Stanford, CA: Stanford University Press.

Weber, M. (2010). *The Protestant Ethic and the Spirit of Capitalism,* ed. and trans. S. Kalberg. New York: Oxford University Press.

Westman, R. and Lindberg, D. (1990). Introduction, in D. Lindberg and R. Westman (eds), *Reappraisals of the Scientific Revolution*. New York: Cambridge University Press, pp. xvii–xxvii.

White, L. (2000a). *Speaking with Vampires: Rumor and History in Colonial Africa*. Berkeley, CA: University of California Press.

White, L. (2000b). Telling More: Lies, Secrets and History. *History and Theory* 39(4): 11–22.

Wilson, T. (1995). Foucault, Genealogy, History. *Philosophy Today* 39(2): 157–70.

Winchester, S. (2008). *The Man Who Loved China: The Fantastic Story of the Eccentric Scientist Who Unlocked the Mysteries of the Middle Kingdom*. New York: HarperCollins.

Wynne, B. (1992). Misunderstood Misunderstanding: Social Identities and Public Uptake of Science. *Public Understanding of Science* 1(3): 281–304.

Zhan, M. (2009). *Other-Worldly: Making Chinese Medicine through Transnational Frames*. Durham, NC: Duke University Press.

Index

Abraham, Itty
  on postcolonial analysis
    151–3
  postcolonial techno-science
    145
  science and caste viii
actor–network theory (ANT)
  colonialism and 115–19,
    142–3
  feminism and 121
  founders of 111
  Latour and 160–1, 164–6
  Law and Lin and 116–17,
    135–6
  Morgan and bush pump
    120–33
  Morita on 115
  spokespersons 149
  Anderson and
    postcolonialism 148–50
Adas, Michael 76
Afghanistan, return of Taliban
  16, 154–5
African countries
  COVID infections 46–8
  scientific revolution 67
Agence France-Presse (AFP),
  vaccination misinformation
  34
Althusser, Louis, interpellation
  and 123–4

*American Journal of Public
  Health* 58
Anderson, Benedict, imagined
  community 55
Anderson, Warwick 27
  actor–network theory 117,
    148–50
  "Asia as a method" 150
  on cultivation of whiteness
    55
  new tools 118
  postcolonial science 32–3,
    144, 145
  "Postcolonial Technoscience"
    22
  response to Abraham
    151–3
animism 115
"Anthropocene" and white
  supremacy 119–20
*Anthropology and the Colonial
  Encounter* (Asad) 24
anti-science
  dualist self/other 33–4
  "experts" invoked in
    29–42
  invoking scientists in
    35–40
  political aspects 31–2
  see also misinformation and
    conspiracy theories; truth

217

226 Index